Palgrave Advances in Possibility Studies

Series Editors
Vlad Glăveanu
Business School
Dublin City University
Dublin, Ireland

Wendy Ross
Psychology
London Metropolitan University
London, UK

Constance de Saint Laurent
Department of Psychology
National University of Ireland, Maynooth
Maynooth, Kildare, Ireland

Palgrave Advances in Possibility Studies explores how humans engage with the possible across multiple domains including futures thinking, embodied and extended cognitive science, educational practices, creative arts, and design. This transdisciplinary series integrates different theoretical frameworks and methodological approaches to understand how possibilities are discovered, created, and enacted across different contexts and systems. Spanning future studies, creativity studies, cognitive philosophy, serendipity research, educational theory, cultural studies, and art/design practice, the series examines possibility-thinking and possibility-making as embodied, distributed processes that emerge through the interaction of minds, bodies, materials, and environments.

The series welcomes proposals for short (25,000-50,000 words) and standard length monographs (80,000+ words), authored or edited.

Nicolas B. Verger • David Mattson
Editors

Possibility Studies and Societal Futures

Imagination, Change and Preservation Across Disciplines

palgrave
macmillan

Editors
Nicolas B. Verger
Centre for Possibility Studies
Dublin City University
Dublin, Ireland

David Mattson
Freddie Reisman Center for
Translational Research in Creativity
and Motivation
Drexel University
Philadelphia, PA, USA

ISSN 3091-2571 ISSN 3091-258X (electronic)
Palgrave Advances in Possibility Studies
ISBN 978-3-032-26378-0 ISBN 978-3-032-26379-7 (eBook)
https://doi.org/10.1007/978-3-032-26379-7

Cover credit: Nastasic

This Palgrave Macmillan imprint is published by the registered company Springer Nature Switzerland AG.
The registered company address is: Gewerbestrasse 11, 6330 Cham, Switzerland

To Vlad P. Glăveanu, for his invitation to serve as co-editor of this book—and for his trust, friendship, and unwavering support during my postdoctoral studies with him. Those two years have passed so quickly.
To all the contributing authors, for their engagement and collaboration throughout the editorial process.
To David and his growing family, to whom I wish many shared moments of joy and peace of mind.
To Laura, for the myriad ways in which we preserve life together.
Nicolas B. Verger

To the Possibility Studies Community, for providing a space unlike anywhere else. Thank you for this opportunity to share in building something truly unique.
To Erin and Lincoln, for their patience and generosity throughout this process.
To Nicolas, for leading this project with grace, positivity, and wisdom.
David Mattson

About the Conference

The 4th International Conference of Possibility Studies took place from 8–12 July 2024 at the University of Cambridge, convened under the theme "Expanding Conceptions for Research on Possibility." Hosted by the Faculty of Education in collaboration with the Possibility Studies Network, the conference brought together hundreds of scholars, researchers, and practitioners from the social sciences, humanities, arts, design, and educational research to interrogate and advance how we understand, study, and cultivate the possible in individual, collective, and cultural life.

At its core, the event centred on interdisciplinary inquiry into the nature and dynamics of possibility—from foundational theoretical questions to practical methods that expand the horizons of inquiry and action. The programme was structured around a rich set of thematic streams, including future-thinking, methodological innovations, shared possibilities, the foundations and power of possibility, digital technologies and AI, creativities research, researching in and with worlds of possibility, the futurities of social enquiry, possibility and culture, and human development. This diversity mirrored the field's commitment to pluralistic and reflexive approaches that treat possibility as both an empirical object and an ontological orientation for research.

Distinguished keynote speakers offered expansive perspectives on the promise and challenges of possibility-oriented scholarship. Contributions from leading figures, including Lord Simon Woolley, Keri Facer, Katina Michael, Nathan Crilly, Hilary Cremin, Hanna Meretoja, Jonas Bozenhard, Eleanor Dare and Penny Hay, stimulated dialogues that bridged theory, practice and policy, and invited participants to rethink assumptions about agency, futures, creativity and social change.

Across paper sessions, workshops and exchanges, the conference showcased cutting-edge research and emerging methodologies that interrogate how individuals and communities envision alternative futures, make sense of uncertainty, and enact transformations. The event not only consolidated the burgeoning identity of Possibility Studies as a field but also fostered dynamic networks and collaborations that will shape future inquiry into the possible across disciplines. The present edited volume is testimony to the intellectual exchanges and collaborations fostered by the 4th International Conference of Possibility Studies, capturing both the depth of ongoing debates and the field's expanding horizons.

Contents

About the Editors

Nicolas B. Verger PhD, is a postdoctoral researcher at the Centre for Possibility Studies, Dublin City University, Ireland. Trained in Applied Psychology at Glasgow Caledonian University in Scotland, he explores the relationships among creativity, resilience, care, and ecological preservation. He conducted his postdoctoral research under the supervision of Professor Vlad P. Glăveanu at Dublin City University (2024–2026), further developing his work at the intersection of creativity, culture, and possibility studies. Verger is the lead author of the creative preservation framework, which advances an axiological turn in creativity research by reorienting creative thinking towards continuity, sustainability, and the protection of socio-ecological relations. His publications appear in interdisciplinary peer-reviewed journals, including *Review of General Psychology*, *Organization*, *The Journal of Creative Behavior*, *Psychology of Aesthetics, Creativity, and the Arts*, *Thinking Skills and Creativity*, and *Possibility Studies & Society*. He serves as Assistant Editor for *Possibility Studies & Society*. He is also co-editor, alongside Prof. Vlad P. Glăveanu and Dr. Constance de Saint Laurent, of the second edition of *The Palgrave Handbook of Creativity and Culture Research*.

David Mattson PhD, served as a postdoctoral fellow with the *Freddie Reisman Center for Translational Research in Creativity and Motivation* (FRC) and the Drexel School of Education from 2023–2025. As a member of the FRC, Mattson assisted in a global research effort to gather valuable insights on creativity and motivation. His efforts to help translate these insights resulted in his contributions to the FRC's publication of

Connecting Creativity and Motivation Research with End-Users: Lab to Learner published by Cambridge University Press. He is a co-author of their publication *Crisis in Teacher Education: Results of Ignoring Studies in Creativity*. Mattson's current research explores teachers' experiences of wonder and creativity inside and outside the classroom. As one of the few researchers around the globe studying wonder in educational contexts, his research explores the nature of wonder as a phenomenon essential to learning new ideas and fostering human flourishing.

List of Contributors

Danilo Audiello PhD, is co-founder and Creative Director working in the field of Science Communication and Illusion on science communication, illusion, creativity, and public engagement with science, Academy of Magic & Science (accelerated at Cambridge Judge Business School, University of Cambridge), Cambridge, United Kingdom.

Pamela Burnard PhD, is Professor of Arts, Creativities and Educations working in the field of Arts and Creativity in Education on the plurality of creativities across education sectors into employment, Faculty of Education, University of Cambridge, Cambridge, United Kingdom.

Laura Colucci-Gray PhD, is Professor of Science and Sustainability Education working in the field of Science and Sustainability Education on participatory methodologies, socio-environmental issues, STEAM education, and garden-based education, Moray House School of Education and Sport, University of Edinburgh, Edinburgh, United Kingdom.

Luis de Miranda PhD, is an associate professor working in the field of Philosophy and Existential Creativity on eudaimonic growth, existential creativity, and philosophical health, Centre for Research Ethics and Bioethics, Uppsala University, Uppsala, Sweden.

Donald Grey PhD, is an emeritus professor working in the field of Science and Sustainability Education on science and sustainability issues, Goethean science, STEAM, and garden-based education, School of Education, University of Aberdeen, Aberdeen, United Kingdom.

Harriet Hand PhD, is a senior research associate working in the field of Graphic Arts, Design and Education on speculative practices, complexity, and processes of change, School of Education, University of Bristol, Bristol, United Kingdom.

Leyya Hoosen PhD, is a lecturer working in the field of Socio-Cultural Digital Anthropology on gender, decoloniality, religion, digital identity, and women in South Africa's Islamic communities, Department of Political Studies, University of the Witwatersrand, Johannesburg, South Africa.

Vicky Hunter PhD, is Professor of Site Dance and visiting research fellow in Dance working in the field of Dance and Performance Research on choreographic exploration of space and place, human-nonhuman connectivity, and embodied movement enquiry, Bath Spa University, Bath, United Kingdom.

Francesca Lorenzi PhD, is an assistant professor working in the field of Philosophy of Education on creativity in education, intergenerational creativity, and dialogical theory and practice, Institute of Education & Centre for Possibility Studies, Dublin City University, Dublin, Ireland.

David Mattson PhD, is a research associate and fellow working in the field of Creativity and Motivation Research on teachers' experiences of wonder and creativity, Freddie Reisman Centre for Translational Research in Creativity and Motivation (FRC) & Drexel School of Education, Drexel University, Philadelphia, United States.

Tim Palmer PhD candidate, is Head of Learning and Teaching working in the field of Music Education on music education and participation, play, improvisation, and conservatoire pedagogies, Guildhall School of Music & Drama, London, United Kingdom.

Karabo-Maya Rodwell is a PhD candidate and doctoral researcher working in the field of Socio-Cultural Anthropology on youth performance theatre, political memory, resistance, art and activism, and Southern African youth, Department of Anthropology, University of the Witwatersrand, Johannesburg, South Africa.

Antonia Symeonidou PhD, is a creative learning educator and entrepreneur working in the field of Science Education and Innovation on science education and wonder-driven innovation, Academy of Magic & Science (accelerated at Cambridge Judge Business School, University of Cambridge), Cambridge Innovation Hub, Cambridge, United Kingdom.

Portia Ungley PhD, is a poet and education researcher working in the field of Creative Education and Wellbeing on poetic inquiry, holistic well-being for creatives, artistic research, and transformative pedagogy, Faculty of Education, University of Cambridge, Cambridge, United Kingdom

Nicolas B. Verger PhD, is a postdoctoral researcher working in the field of Applied Psychology on creativity and creative preservation, Centre for Possibility Studies, Dublin City University, Dublin, Ireland.

List of Figures

List of Pictures

List of Tables

CHAPTER 1

Possibility Studies and Societal Futures: An Introduction

Nicolas B. Verger and David Mattson

Introduction

The future is behind us. In the long accumulation of what humans experience, imagine, and create, societal futures can only be realised through what came before. Of course, that is partly why we think of children as the future of societies: they use and build upon the cultural legacy of their elders. Infants cannot create from a blank slate, nor can the future be conceived *ex nihilo*. We inherit actualised societal futures through the affordances we prolong. Thus, the desire to create by destroying structures to replace them—the *modus operandi* of Westernised societies since the modern era (Aghion & Festré, 2017; Schumpeter, 1962)—seems to reflect a profound misunderstanding of how societies create the conditions for

N. B. Verger (✉)
Centre for Possibility Studies, Dublin City University, Dublin, Ireland
e-mail: Nicolas.verger@dcu.ie

D. Mattson
Freddie Reisman Center for Translational Research in Creativity and Motivation, Drexel University, Philadelphia, PA, USA
e-mail: dmm563@drexel.edu

N. B. Verger, D. Mattson (eds.), *Possibility Studies and Societal Futures*, Palgrave Advances in Possibility Studies,
https://doi.org/10.1007/978-3-032-26379-7_1

future, sustainable life. The future is behind us, because what our societies become shall depend on what the youngest forms of life make of the oldest forms of imagination.

The Ancient Greeks intuited that the future of the society they inherited was made possible only through both change and preservation. Many readers will be acquainted with the myth of Prometheus (literally, "foresight", that is, the one looking towards what is not yet there) and his brother Epimetheus (namely, "hindsight", that is, the one looking at what is already there). As the tale has it, the two brothers were Titans, original forces of the universe and thus greater than the gods themselves, who joined Zeus in fighting the other Titans. After the war was won, Zeus asked the two brothers to distribute parts of the gods' attributes among living creatures. Epimetheus gave non-human animals strength, speed, wings, and protective features, but did not anticipate that no gifts would be left for humans. Humanity was left weak and defenceless against other animals.

Prometheus, realising this oversight as he cared for the humans he himself had moulded from clay, resolved to compensate for what Epimetheus had forgotten. He stole fire from Hephaestus, the god of blacksmithing and fire, and delivered it to humankind. Humans were granted the seed of technē—craft, calculation, and the ability to create. With fire, they forged weapons, built cities, travelled the seas, waged war, and built empires. Eventually, humans rebelled against the very elements they were meant to worship and, with the development of civilisation, came to regard themselves as gods. Zeus punished Prometheus by binding him to perpetual suffering, chaining him to a cliff with unbreakable bonds. Each morning, a vulture would eat his liver. Each night, his liver would regrow for the bird to peck and claw it the next day.

This myth has multiple layers of meaning and moral implications. We cannot explore them all in this introduction. But the most common is certainly that foresight must answer to hindsight, and vice versa. Innovation must remain tethered to memory and continuity with the ecological milieu that conditions one's birth: humans should always have remained tethered to Zeus and to the animals with divine qualities (and thus partly divine as well). But the story also tells us that Prometheus, who moulded humanity with his hands, as a potter would do, cared for the fruit of his creation, and that caring for what we create is as old as humanity itself. The myth suggests that societies often seek growth by erasing what came before, but such actions invite negative consequences. When our creations disrupt and disregard the past, they ultimately undermine sustainable futures. Zeus punished Prometheus not because humans had the power to create, but because this creative force led them to a hubris that made humans turn their backs on the divinities. One can replace "divinities" with "ecological milieu" to see that this intuition may be warranted today. The future of societies depends on how we imagine and create in continuity and care for what already exists, and has made our lives possible.

Societal futures have attracted work from various angles. From reflections on post-growth (Savini, 2023) to civilisational collapse (Brozović, 2023), from sustainability goals (Arora-Jonsson, 2023) to the effects of artificial intelligence (Sharma, 2024), Futures Studies extend the Greek intuition with contemporary science. As they form, communities cannot live without the interdependence they establish with diverse ecological systems (Oman-Reagan, 2019). Following the disruption of the COVID-19 pandemic and the questioning it imposed on societies' futures, another research field has emerged: Possibility Studies (Glăveanu, 2023; Ross et al., 2023).

A distinctive feature of Possibility Studies, which complements the knowledge within Future Studies, is its focus on the cognitive, conative, and behavioural processes that explain how individuals and communities explore possibilities in the past, present, near, and distant future. This transdisciplinary science examines individuals and groups through an embodied, relational framework, investigating how they co-create and navigate ecological, socio-cultural, and political worlds (Escobar, 2023). Possibility Studies address the fundamental questions of societal becoming from exploring different perspectives from a variety of disciplines, including social psychology, education, ethics, aesthetics, cognitive science, developmental studies, cultural studies, and behavioural economics, to name a few (Beghetto, 2023; Beyer, 2025; Copeland et al., 2023; Glăveanu & de Saint Laurent, 2025; Ross, 2023; Verger, 2025). The transdisciplinary formation of Possibility Studies is warranted by the intricate themes it seeks to study.

Imagination, Change, and Preservation Across Disciplines

The societal futures we imagine are never conjured in a vacuum—they are shaped by the institutions we inhabit, the stories we are told, and the possibilities we preserve. First among these institutions is education, which does far more than transmit knowledge; it determines what is considered worth knowing, and therefore what future generations believe is worth changing and preserving. The arts also remain a crucible of societal imagination, as they provide communities with outlets for new ecologies of creative imagination for social change (Glaveanu & de Saint Laurent, 2025; Glăveanu & Clapp, 2017). Through art, communities process their grief, celebrate their hopes, and engage with different visions of the world. But perhaps the deepest catalyst of all is wonder, the soulful stirring that moves a child to ask why the sky is dark when the universe is full of stars, or what it means to live a good life (Glăveanu, 2020; Mattson, 2023). Wonder is the beginning of philosophy, through which societies have long interrogated their own assumptions, resurrected forgotten possibilities, and imagined different ways of being together. A society that starves wonder loses more than its artists and philosophers. It loses its capacity to imagine itself otherwise, and with it, the societal futures that might have been.

This book is an interdisciplinary anthology exploring possibility studies and societal futures. Emerging from the 4th Possibility Studies Conference, which was held at the University of Cambridge between 8–12 July 2024, this work brings together a diverse group of researchers who draw on their expertise to examine how our shared futures may depend on imagination, change, and preservation. This volume brings together ten original contributions spanning creativity research and social innovation; applied philosophy (including bioethics); philosophy of education and applied education; the performing arts (particularly choreography); political studies; socio-cultural anthropology; poetry; and epistemology. Each chapter emerges from a discipline shaped by the institutions as they are—by the conventions, constraints, and inherited assumptions of its field—yet, each takes a path less travelled.

Creative Preservation as a Unifying Framework

A thread running through the book centres on a common framework. It is the concept of creative preservation (Verger, 2024; Verger et al., 2024, 2025). Creative preservation refers to forms of creative practice that sustain existing relationships and enrich them, without diminishing their potential for future configurations. Rather than pursuing constant novelty or endless growth, the concept opens up analyses about the actors and practices that care for, repair, and reconfigure the present in ways that keep possibilities open. Creative preservation involves using creative thinking to maintain the continuity of what people hold dear. More formally, creative preservation is defined as "practices of creation that prevent the decay of existing materials and ideas by updating and adapting them, or re-expressing them in another way through the exploration of their affordances" (Verger et al., 2024, p. 270).

Hence, creative preservation reframes preservation not as passive conservation or resistance to change, but as an active and imaginative process. It recognises that materials, practices, and ideas inevitably transform over time; the question is not whether change occurs, but how it unfolds in continuity with what conditioned its emergence. Creative preservation, therefore, privileges forms of transformation that extend the life, relevance, and relational richness of what already exists. Creative preservation, for instance, encompasses practices such as repair, maintenance, upcycling, low-tech innovation, craft for social justice, and bricolage as generative responses to ecological and social concerns. At its core, creative preservation is relational and temporal. It seeks to strengthen bonds among people, materials, and environments, while cultivating awareness of the links among past, present, and future. By exploring the affordances of existing materials and situations, practitioners reveal latent possibilities and activate new configurations without exhausting the conditions for further change. Each chapter offers examples of the application of creative preservation as an analytical tool to deepen understanding of the relationship between imagination, change, and preservation.

Structure of the Book

This book is intended to be read from front to back. Of course, each chapter may be read individually, but the strength of the volume lies in the plurality of voices it gathers across disciplines. The journey begins at scale (at the level of global social and ecological systems) by examining the possibility that innovation may, under certain conditions, function not as a force of disruption but of preservation. It then turns to questions concerning our regard for the young and the old, before moving to methodological considerations and reflections on academic practice in relation to societal change. The book concludes by opening onto the educative, and thus societally transformative, potential of the stupor and wonder of what has come before us and is just there, at hand.

This opening chapter, *Possibility Studies and Societal Futures: An Introduction*, has situated the volume within the broader transdisciplinary landscape of Possibility Studies. It has articulated the need to comprehend the relationships between imagination, change, and preservation for societal change, and introduced creative preservation as a conceptual lens through which this tension may be re-examined.

In Chap. 2, *Social-(Ecological) Innovation as a Scaled Form of Creative Preservation*, Nicolas B. Verger examines whether social–(ecological) innovation can be understood as a scaled form of creative preservation oriented towards societal futures. Through a corpus analysis of 249 articles across leading innovation journals, this work reveals the marginality of social–ecological concerns within mainstream innovation discourse. The chapter proposes that creative preservation operates through anticipatory stewardship, intergenerational transmission, affective infrastructure, and the political negotiation of futures. In doing so, it reframes social-(ecological) innovation not as the pursuit of novelty alone, but as a reflexive, collective practice of care aimed at sustaining the relational, material, and affective conditions from which meaningful societal futures can be composed.

In Chap. 3, *Crealectics: A Method for Compossible Worlds*, Luis de Miranda diagnoses a contemporary "compossibility crisis". He proposes what he labels "crealectics" as a structured process for ensuring that the possibilities we generate can coexist rather than be mutually destructive. The chapter outlines a method for advancing towards compossible worlds, and opens up a relational, process philosophy of intercreation. This work contributes to the analysis of how societies are always created within systems that reciprocally create them. It proposes that we think of innovation in light of a "Regenerocene": an era not only focused on harm reduction but on actively restoring and regenerating the ecological, social, and cultural systems that industrial civilisation has depleted.

In Chap. 4, *Can an Old Dog Teach New Tricks? Older People as Agents of "Slow Futurition"*, Francesca Lorenzi reframes our assumptions about age, temporality, and futurity. Against narratives that equate innovation with youth, it foregrounds older adults as custodians of long memory and as active agents of societal becoming. This chapter shows that preservation is not inertia but a condition of meaningful continuity, and that "slow futurition" becomes a counterpoint to accelerationist imaginaries. This work opens up novel considerations about the potential and dignity of older adulthood to foster intergenerational continuity while cultivating the capacity to engage critically and creatively with inherited knowledge.

Having engaged with global innovation systems and intergenerational continuity, the journey turns towards research itself. In Chap. 5, *Expanding Possibilities for Research Methodologies that Inspire Change: Why 'Thing-Power' Matters*, Pamela Burnard, Laura Colucci-Gray, Donald Gray, and Tim Palmer interrogate methodological assumptions for societal futures. Drawing attention to the vitality of material things, the chapter expands the scope of agency beyond the human and invites researchers to reconsider how objects participate in the co-creation of futures. Drawing on three case studies, it shows how school gardening practices, mathematical art practices, and musical play practices open up assemblages of human and non-human actors, each offering distinct affordances for investigating societal change through educative practices.

In Chap. 6, *Rules for Researching: Rules for Possibility*, Harriet Hand continues this reflexive turn by examining the implicit norms governing academic inquiry. If possibility is constrained by what research methods allow us to see, then being creative about the rules we establish as constraints enables us to endure and preserve our capacity to continue engaging in the practices we care about—in this case, scientific inquiry. Rules, in their varying degrees of rigidity, can function as instruments of change, and imagining new rules entails opening up certain possibilities while constraining others. The chapter asks how academic conventions might both enable and foreclose societal futures.

In Chap. 7, *Dances Yet to Come: The Possibilities of Performance that Never Was—or Only Ever Lived in Our Imaginations*, Vicky Hunter reflects upon the COVID-19 pandemic, which disrupted the rules of universities and, with them, the conditions under which students and professors learned and taught choreography. The rupture in the continuity of academic processes—typically structured around measurability, outcomes, and end products—drew attention to alternative modes of relating to dance practice. The chapter shows that this displacement of rules reoriented engagement towards aspects of practice that had faded from perception, allowing them to regain value. It opened up different ways of valuing process, creativity, experimentation, and the material significance of elements that had previously been overlooked, suddenly revealing new affordances and meanings about how we engage with what we know will never be.

In Chap. 8, *Troubling the Recovering Academic: Poetry as Material Resistance*, Portia Ungley further unsettles the institutional self-image of academia. Through poetry, she explores how language itself can resist the meritocracy of production inherent in academic culture and preserve spaces for vulnerability, ambiguity, and critique. She shows that academia is not merely a producer of outputs but a situated body navigating institutional pressures. Poetry becomes an aesthetic way of knowing and engaging with the world, beyond data gathered through predominantly positivist and hegemonic modes of inquiry. It offers a mode of resistance to institutional rigidities that silence the voices of those who sense and experience the world differently.

In Chap. 9, *The Mythologies of Possibility: Global South, Care and Decolonial Scholarship*, Leyya Hoosen and Karabo-Maya Rodwell expand this critical discussion to geographical and epistemic frame. This chapter interrogates whose possibilities are legitimised within dominant academic narratives and foregrounds care as a decolonial practice. The authors examine the limits of transposing models of possibility from the Global North to the Global South. They argue that language and its articulation shape the ways in which societal change can be imagined and enacted, and prompt the critical question: societal change for whom?

The journey culminates in two chapters that open this book onto the question of how to cultivate the conditions for societal change. In Chap. 10, *The Vitality of 'Stuporism'*, Antonia Symeonidou, Danilo Audiello, and Pamela Burnard introduce the practice of stupor (stuporism) as the cultivation of a generative rupture in habitual perception. Rather than celebrating constant activity and anaesthetic productivity, which render stimuli imperceptible in the background of experience, this work reclaims astonishment and perceptual disorientation as catalysts for transformative scientific and educational inquiry. The authors propose a six-step framework that moves from encountering phenomena to questioning and ultimately to enacting change. They provide concrete pedagogical design examples drawn from the art of magic, positioning stuporism as an epistemic catalyst.

Finally, in Chap. 11, *Wonder, Education, and Possible Futures*, David Mattson critically reviews existing definitions of wonder, highlighting the absence of a unifying account of the phenomenon. He identifies four categories into which these definitions fall and explores wonder as a disciplined openness that grants unfolding events a dignity often overlooked in the stream of everyday life. Hence, this book closes with the consideration that the future depends on how we attend to and give worth to what already is. The future is before us.

References

Aghion, P., & Festré, A. (2017). Schumpeterian growth theory, Schumpeter, and growth policy design. *Journal of Evolutionary Economics, 27*(1), 25–42. https://doi.org/10.1007/s00191-016-0465-5

Arora-Jonsson, S. (2023). The sustainable development goals: A universalist promise for the future. *Futures, 146*, 103087. https://doi.org/10.1016/j.futures.2022.103087

Beghetto, R. A. (2023). Broadening horizons of the possible in education. *Possibility Studies & Society, 1*(4), 414–426. https://doi.org/10.1177/27538699231182014

Beyer, M. (2025). Imagine the possible! Dealing with barriers to new economic thinking in higher education. *Possibility Studies & Society*, 27538699251381964. https://doi.org/10.1177/27538699251381964

Brozović, D. (2023). Societal collapse: A literature review. *Futures, 145*, 103075. https://doi.org/10.1016/j.futures.2022.103075

Copeland, S., Arfini, S., & Ross, W. (2023). Enacting chance and the space of possibilities. *Possibility Studies & Society, 1*(3), 263–268. https://doi.org/10.1177/27538699231191006

Escobar, A. (2023). Welcome to possibility studies. *Possibility Studies & Society, 1*(1–2), 56–59. https://doi.org/10.1177/27538699231171450

Glăveanu, V. P. (2020). Wonder : The Extraordinary Power of an Ordinary Experience. Bloomsbury Academic.

Glăveanu, V. P., & Clapp, E. P. (2017). Distributed and participatory creativity as a form of cultural empowerment: The role of alterity, difference and collaboration. In Alterity, values, and socialization: Human development within educational contexts (pp. 51-63). Cham: Springer International Publishing.

Glăveanu, V., & de Saint Laurent, C. (2025). Methods as horizons of possibility: HOPE for the future. *Possibility Studies & Society, 3*(2), 179–183. https://doi.org/10.1177/27538699251343157

Glăveanu, V. P. (2023). Possibility studies: A manifesto. *Possibility Studies & Society, 1*(1–2), 3–8. https://doi.org/10.1177/27538699221127580

Mattson, D. (2023). A Positionality of Wonder: Connecting Teachers' Creativity Inside and Outside the Classroom. Drexel University.

Oman-Reagan, M. P. (2019). Politics of planetary reproduction and the children of other worlds. *Futures, Human Colonization of Other Worlds, 110*, 19–23. https://doi.org/10.1016/j.futures.2019.02.015

Ross, W. (2023). Embracing difference and cultivating hope: The transformative potential of possibility studies. *Possibility Studies & Society, 1*(1–2), 243–259. https://doi.org/10.1177/27538699231172436

Ross, W., Glăveanu, V., & Baumeister, R. F. (2023). The new science of possibility. *Possibility Studies & Society, 1*(4), 399–403. https://doi.org/10.1177/27538699231211629
Savini, F. (2023). Futures of the social metabolism: Degrowth, circular economy and the value of waste. *Futures, 150*, 103180. https://doi.org/10.1016/j.futures.2023.103180
Schumpeter, J. A. (1962). *Capitalism, socialism, and democracy* (3rd ed.). Harper Perennial.
Sharma, S. (2024). Benefits or concerns of AI: A multistakeholder responsibility. *Futures, 157*, 103328. https://doi.org/10.1016/j.futures.2024.103328
Verger, N. B. (2024). Transformational creativity, care, and the common good: Toward a refined definition? *The Journal of Creative Behavior*, jocb.1511. https://doi.org/10.1002/jocb.1511
Verger, N. B. (2025). The aesthetics of the possible. *Possibility Studies & Society*, 27538699251318068. https://doi.org/10.1177/27538699251318068
Verger, N. B., Duymedjian, R., Wegener, C., & Glăveanu, V. P. (2024). Creative preservation: A framework of creativity in support of degrowth. *Review of General Psychology, 28*(3), 268–281. https://doi.org/10.1177/10892680241256312
Verger, N. B., Duymedjian, R., & Glăveanu, V. P. (2025). The meritocracy of preservation: Reimagining merit beyond production. *Organization, 33*(2), 294-312.

CHAPTER 2

Social–(Ecological) Innovation as a Scaled Form of Creative Preservation for Habitability

Nicolas B. Verger

Introduction

Innovation is centrally concerned with change. As a central instrument of economic growth, innovation appears to some as a 'force for creating a better world' (Schot & Steinmueller, 2018, p. 1561). Yet the scientific and technological—and thus societal—implications of innovation often follow a logic of resource consumption, under a so-called Schumpeterian paradigm of creative destruction and radical change in market structures (Schumpeter, 1962; Tidd & Bessant, 2021). The form of innovation that is perhaps most easily reconcilable with the axiological outlook proposed by Schot and Steinmueller (2018) is social innovation. This replaces the goals of economic growth with the aims of social justice, or at least seeks to create social value alongside economic sustainability. Historically, for instance, novel ideas have been implemented to drive societal change,

N. B. Verger (✉)
Centre for Possibility Studies, Dublin City University, Dublin, Ireland
e-mail: Nicolas.verger@dcu.ie

N. B. Verger, D. Mattson (eds.), *Possibility Studies and Societal Futures*, Palgrave Advances in Possibility Studies,
https://doi.org/10.1007/978-3-032-26379-7_2

giving rise to institutions such as national health services, kindergartens, or egalitarian voting systems (Luria & Kaufman, 2025; Mulgan, 2006, 2012). We cannot say that these social innovations are independent of the economic system that gave them birth and sustains them, but profitability is not their main goal. More formally, social innovation can be defined as 'innovative activities and services that are motivated by the goal of meeting a social need and that are predominantly diffused through organisations whose primary purposes are social' (Mulgan, 2006, p. 146).

Driven by the goal of addressing societal issues for a better future, social innovation starts from the premise that 'for-profit enterprises need to take into account not only the interest of owners but also the interest of other stakeholders *that can affect or be affected* by the activity of a for-profit enterprise' (Phillips et al., 2014, p. 2, my emphasis). For it to be implemented, social innovation requires a reflexive, caring stance—one that considers how one's actions may affect or be affected by others. From my reading, this stance entails creating practices and structures that prevent disruption and rupture or actively contribute to people's worth and dignity. It does so by enhancing their capacity to continue living, improving their lives, and ensuring a sustainable future. This seems to qualify, at least in part, social innovation as a large-scale creative preservation practice. I am tempted to suggest that social innovation, at scale, is a form of innovation for the preservation of social-ecological futures. This chapter explores the nuances of this claim.

Creative preservation can be defined as the use of creativity to explore the affordances of the elements that constitute one's socio-ecological milieu while ensuring that such exploration preserves the possibility for their future reconfiguration and rearrangement (Verger, Duymedjian, et al., 2024). Creative preservation consists of the use of creativity as a means of preservation, aimed at preserving the affordances of existing materials and ideas from decay or actively aiming at promoting their continuity. Practically, this should be achieved by updating, adapting, or re-expressing elements beyond their original or intended design, and by using, connecting, and arranging them in novel ways (Glăveanu, 2012; Wegener, 2023).

So far, creative preservation has been discussed in the following terms. Researchers have conceptualised the inherent creativity of marginal practices such as upcycling, bricolage, low-tech approaches, or certain uses of craft for social justice (Verger, Duymedjian, et al., 2024). In design studies, Le Masson et al. (2026) have argued that the preservation of structural elements is pivotal for creativity and that one should preserve the fundamental structures that allow the deployment of certain forms of innovation. Creative preservation might therefore be tentatively understood as operating through two interrelated pathways. On the one hand, preservation may sustain the structural conditions that make creativity possible. On the other hand, through the exploration of new affordances, new uses and values of what already exists, creativity could generate renewed worth in those elements. According to Verger, Duymedjian, (2024), this renewed worth would provide a rationale for their continued preservation. Creative preservation has also been discussed in accounting for how some organisational processes discard creative preservation practices—considered marginal to the core meritocratic/economic system—despite their pivotal role in ensuring cohabitation and collective robustness through maintenance, repair, and continuity (Verger, Duymedjian, et al., 2025). This emerging body of work indicates that creativity focused on care and dignity still remains marginal in academic analyses and organisational focus (Verger, 2024).

A limitation of this work on creative preservation has been the lack of empirical studies exploring the hypothesis proposed by Verger, Duymedjian, et al. (2025). Some researchers are beginning to contribute to addressing this issue (Le Masson et al., 2026). However, although socially promising, this endeavour remains notably overlooked. In what they analysed as a 'Meritocracy of Production', Verger, Duymedjian, et al. (2025) proposed that organisational narratives influence how collectives perceive the value of disruptive novelty as emancipatory, rather than valuing the ongoing exploration of existing possibilities to preserve conditions for habitability. This latter point—a form of creativity that ensures the affordances which contribute to habitability can continue to unfold—encapsulates the essence of creative preservation.

Social innovation precisely aims to ensure habitability, since this large-scale creative enactment focuses on quality of life, social justice, collective well-being, and, globally, thus ensuring meaningful and sustainable living conditions. Concerned that mainstream innovation research has overlooked how innovation can modify and disrupt society by eroding established socio-ecological relationships, a growing number of researchers have updated the notion of social innovation. They now call for greater attention to social–ecological innovation (Dennis et al., 2016; Dennis & James, 2018; Edgeman & Eskildsen, 2014; Melnykovych et al., 2018; Olsson & Galaz, 2012). This emerging literature remains marginal, despite offering examples of innovation for the preservation of social–ecological futures.

Aligned with the proposition that this marginality may be accounted for by narratives and discourses that give limited consideration to values of preservation (Verger, Duymedjian, et al., 2025), this chapter presents the results of a corpus analysis of ten major innovation journals, covering 249 studies in the field. This review contributes to answering three questions:

1. *What is the percentage of social–(ecological) innovation in this body of work?*
2. *How do innovation studies inform different applications of the notions of preservation and conservation?*
3. *When does social–(ecological) innovation qualify as a scaled form of creative preservation?*

This third question is the pivotal contribution of this chapter to creative preservation research.

Methods

Across the ten innovation journals listed in ABS 2024, I searched for the term preservation using the journal search tab. I extracted articles containing at least one occurrence of the string preserv*, with an arbitrary cap of 15 articles per outlet. I then applied the same search strategy using the term conservation, again aiming to include up to 15 articles per outlet. These searches were conducted between 14 and 17 June 2025.

When possible in journal options, I used Boolean operators (preserv* AND/OR conserv*). After verifying the occurrence of these terms in the full manuscripts—and ensuring that they referred to relevant concepts (e.g., excluding uses such as 'preserve the quality of a statistical model' or 'preserve the anonymity of participants')—I downloaded the full PDF versions from the journal websites and conducted verification searches for the terms preservation and conservation to ensure they were relevant articles. Articles containing both terms were included and were not treated as duplicates in the analysis, as research in preservation and conservation studies indicates that authors may use these terms interchangeably (Matthes, 2025). The final sample included empirical articles and review articles, but excluded book reviews. The terms conservatism and conservative were excluded from the analysis.

Textual and Corpus Analysis Using AntConc

To identify word families, collocations, and recurring lexical patterns within the selected corpus, I employed AntConc (v. 4.3.1; Anthony, 2005), a free and widely recognised corpus analysis toolkit. AntConc enabled me to conduct a systematic search for specific terms and their morphological variants (e.g. pre/conserve, pre/conserving, pre/conservation, pre/conserved) across the full text. In addition, I used the software's collocation and concordance tools to examine the contexts in which target words appeared. This facilitated the identification of underlying themes, relationships, and language patterns relevant to my research focus.

I identified word families by configuring search parameters to capture word stems, ensuring comprehensive retrieval of all lexical forms related to the key terms. I employed the collocates function to detect the proximity of associated terms, identifying semantic fields and thematic linkages relevant to the preservation discourse. The corpus I used in this study is freely available: https://osf.io/em8ys/overview. Although commercial software such as NVivo offers robust text-analysis capabilities, I preferred AntConc for several reasons. First, AntConc is specifically designed for linguistic pattern analysis rather than broader qualitative coding, providing more transparent and replicable procedures for concordance and collocation analysis. Second, its lightweight design and flexibility allowed me to conduct rapid iterations of keyword and phrase searches without the extensive project setup required by NVivo. Third, its focus on raw textual data minimised interpretative bias during the initial stages of word identification.

After removing duplicates, I uploaded 251 PDF files into AntConc to construct the corpus. Two files could not be processed by the software. Hence, in total, I conducted the analyses across 249 individual articles (see Fig. 2.1).

I conducted frequency counts and classifications for Sort to Right and Sort to Left. This allowed me to analyse left and right contexts around 'social innovation', 'social-ecological innovation', 'conserv', and 'preserv' to provide an exhaustive frequency count of how articles mobilise, respectively, the notions of social innovation, social-ecological innovation, conservation, and preservation. Finally, I conducted *n*-gram searches (six words following the target term) for the following lexical items:

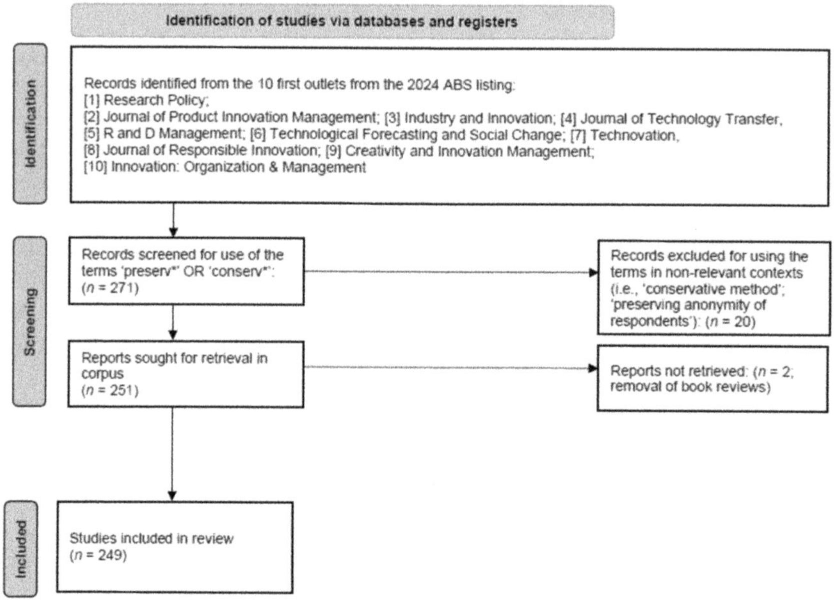

Fig. 2.1 A summary of the search procedure and inclusion of retrieved articles in corpus

Social–(ecological) innovation

- *Social innovation*
- *Social-ecological innovation*
- *Social ecological innovation*
- *Socialecological innovation*

Preservation-related terms (with a wildcard indicating any word preceding *preservation*):

- *preservation of*
- *preserved*
- *preserve*
- *preserves*
- *preserving*

Conservation-related terms (with a wildcard indicating any word preceding *conservation*):

- *conservation of*
- *conserved*
- *conserve*
- *conserves*
- *conserving*

Results

Country Distribution (Based on Lead Author's Institution)

A frequency analysis of the 249 country mentions in my sample reveals a distinctly skewed geographic distribution (see Table 2.1). The United States appears most frequently (n = 28; 11.2% of all references), followed closely by China and Italy (26 mentions each; 10.4%), and then Germany (n = 24; 9.6%). The United Kingdom ranks fifth (n = 19; 7.6%), with the Netherlands (n = 13; 5.2%), France (n = 12; 4.8%), and both Spain and Australia (n = 11 each; 4.4%) forming the next tier, while Sweden contributes ten occurrences (4.0%).

Collectively, these ten countries account for approximately 72% of the dataset, underscoring a pronounced concentration of activity within a relatively small group of nations. The remaining 29 countries each register fewer than ten mentions, including Canada (7), Austria (6), Denmark and India (5 each), Switzerland, Ireland, and Greece (4 each), 5 countries with two mentions, and a further 12 countries represented only once. This pattern highlights the long-tailed nature of the distribution and the comparative scarcity of references to many regions of the world.

What Are the Percentage of Social–(Ecological) Innovation in This Body of Work?

The term *social innovation(s)* appeared 329 times in the corpus, whereas *social-ecological innovation* did not appear at all, nor did the term *social-ecological.* By comparison, the word *innovation* appeared 15,910 times. References to social-ecological innovation, therefore, account for roughly 2% of all occurrences of the word *innovation.*

Beyond merely quantifying textual occurrences, I extended my analysis to the themes of each paper and examined the 249 articles that comprised

Table 2.1 Description of the included studies

Country of lead author	*n (%)*
United States	28 (11.2)
China	26 (10.4)
Italy	26 (10.4)
Germany	24 (9.6)
United Kingdom	19 (7.6)
The Netherlands	13 (5.2)
France	12 (4.8)
Australia	11 (4.4)
Spain	11 (4.4)
Sweden	10 (4.0)
Canada	7 (2.8)
Austria	6 (2.4)
Denmark	5 (2.0)
India	5 (2.0)
Switzerland	4 (1.6)
Ireland	4 (1.6)
Greece	4 (1.6)
Finland	3 (1.2)
New Zealand	3 (1.2)
Norway	3 (1.2)
South Korea	3 (1.2)
Pakistan	2 (0.8)
Portugal	2 (0.8)
South Africa	2 (0.8)
Japan	2 (0.8)
Nigeria	2 (0.8)
Belgium	1 (0.4)
Brazil	1 (0.4)
Colombia	1 (0.4)
Hong Kong	1 (0.4)
Iran	1 (0.4)
Morocco	1 (0.4)
Romania	1 (0.4)
Russia	1 (0.4)
Slovenia	1 (0.4)
Taiwan	1 (0.4)
Turkey	1 (0.4)
Zimbabwe	1 (0.4)

the present sample. I identified 12 articles addressing themes such as social innovation, social–ecological goals, responsible innovation, eco-innovation, social justice, and related topics (Bosisio et al., 2024; Huguenin & Jeannerat, 2017; Kuzma et al., 2018; Lozano & Monsonís-Payá, 2020; Mongelli & Rullani, 2017; Nakagawa & Kosaka, 2022; Roysen et al., 2024; Sauermann et al., 2020; Steinfield & Holt, 2019; Vassallo et al., 2023; Venugopal & Viswanathan, 2019; Villalba Morales et al., 2023).

This accounted for 4.8% of the total sample of included studies. Clearly, these themes are marginal relative to dominant business-oriented innovation foci.

How Do Innovation Studies Inform Different Applications of the Notions of Preservation and Conservation?

Table 2.2 shows an important disparity in the frequency of conservation- and preservation-related terminology. The term *conservation* (1117 occurrences) appears nearly three times more often than *preservation* (396 occurrences). Within their respective word families, *conservation* accounts for 92% of all terms containing the root *conserv**, whereas *preservation* represents only 51% of terms derived from *preserv**. Other forms within the *preserv** family include the verb *preserve* (23%) and the gerund *preserving* (17%). These patterns suggest that conservation is predominantly articulated in a noun-based, relatively passive form, emphasising a state or condition of being conserved. In contrast, preservation is more frequently framed in terms of verbs and gerunds, highlighting active, ongoing

Table 2.2 Occurrences derived from 'preservation' and 'conservation'

Preservation		*Conservation*	
Word	*Occurrence*	*Word*	*Occurrence*
Preserv*	768	Conserv*	1216
Preservation	396	Conservation	1117
Preserve	179	Conserve	49
Preserving	135	Conserving	41
Preserved	41	Conserved	7
Preserves	14	Conserves	2
Preservers	2	–	
Preservatives	1	–	

Note. Occurrences of 'conservatism' and 'conservative' ($n = 153$) were excluded from these analyses processes aimed at maintaining or protecting valued objects or practices. This observation leads to the next stage of the analysis, which examines what innovation researchers perceive as worth conserving and preserving.

Table 2.2 shows the results of the *n*-gram corpus analysis for the wildcard searches *conserv** and *preserv**. For comparative purposes, I report both the total number of occurrences and the thematic distribution of each term across mutually exclusive domains. As shown in Table 2.2, *conserv** (1216 hits) occurs substantially more frequently than *preserv** (768 hits).

Conservation is predominantly used to refer to environmental and ecological safeguarding, including biodiversity, ecosystems, nature, wildlife, land, marine environments, climate, and habitats. This domain accounts for 46% of all conservation-related occurrences. In addition, conservation is used in approximately 28% of cases regarding resources and energy, confirming its association with material and technical forms of resource management. By contrast, conservation is used only marginally to refer to organisational matters: references to safeguarding organisational structures account for just 5% of occurrences. Conservation is also rarely associated with cultural heritage (2%) or social and human values (1%).

Preservation is also used in relation to environmental and ecological domains, though to a markedly lesser extent (25.5%). Beyond this, preservation is strongly associated with organisational and institutional structures, accounting for around 22% of occurrences (institutions, routines, regulation, industrial organisation, status quo, and structures). Preservation further features prominently in relation to family firms and socioemotional wealth (SEW) (19%), reflecting concerns with legacy, identity, and continuity. It is also frequently linked to the maintenance of knowledge, skills, and capabilities (13%), as well as to cultural heritage—including monuments, architecture, traditions, crafts, and collective identity—representing approximately 7% of occurrences. Associations with social and human values remain limited (4.3%).

This pattern suggests that *conservation* and *preservation* are sometimes used interchangeably when researchers discuss ecological contexts, but that they are conceptually distinct in the way innovation researchers use them (see Table 2.3). Both terms refer to the ongoing life of things, people, structures, and so forth. Conservation is primarily used to denote the management and safeguarding of external resources that support organisational activity. Preservation is more closely associated with the protection of an organisation's internal structures, identities, and forms of continuity.

Table 2.3 Frequencies of words associated with conservation and preservation in a (6)*n*-gram corpus search

Conserv (1216 hits)*			*Preserv* (768 hits)*		
Domain/theme of conservation	*Occurrences*	*Percentage*	*Domain/theme of preservation*	*Occurrences*	*Percentage*
Environmental & ecological conservation *(biodiversity, ecosystems, nature, wildlife, land, marine, climate, habitat)*	559	46.0%	**Environmental & ecological preservation** *(environment, biodiversity, ecosystems, natural resources, land, marine areas, wildlife)*	196	25.5%
Resource & energy conservation *(energy, water, soil, materials, recycling, efficiency, agriculture, tillage)*	334	27.5%	**Organisational/institutional structures** *(institutions, routines, regulation, industrial organization, status quo)*	164	21.4%
Psychological/behavioral conservation (COR theory) *(conservation of resources theory, stress, coping, motivation, behavioural models)*	182	15.0%	**Family firms & socioemotional wealth (SEW)** *(family control, legacy, identity, continuity, family wealth, non-financial goals)*	146	19.0%
Organizational/institutional conservation *(institutions, regimes, policy stability, organisational arrangements, status quo)*	61	5.0%	**Knowledge, skills & capability preservation** *(knowledge, know-how, learning, science potential, organisational memory, expertise)*	100	13.0%
Economic/business-related conservation *(firm resources, competitiveness, productivity, cost reduction, labor markets)*	43	3.5%	**Cultural & heritage preservation** *(cultural heritage, monuments, architecture, traditions, crafts, collective identity)*	51	6.6%

(continued)

Table 2.3 (continued)

Conserv (1216 hits)*			*Preserv* (768 hits)*		
Domain/theme of conservation	*Occurrences*	*Percentage*	*Domain/theme of preservation*	*Occurrences*	*Percentage*
Cultural & heritage conservation *(cultural heritage, biocultural diversity, indigenous practices, traditions)*	24	2.0%	**Economic/business asset preservation** *(firm survival, value creation, competitive position, intellectual property, patents, resources)*	48	6.2%
Social & human values conservation *(livelihoods, jobs, human rights, community wellbeing)*	12	1.0%	**Social & human values preservation** *(jobs, wellbeing, human life, freedom, rights, social identity, social cohesion)*	33	4.3%
—			**Data/digital/technical preservation** *(data, digital records, software, evidence, protocols, technological artefacts)*	30	4.0%

When Does Social–(Ecological) Innovation Qualify as a Scaled Form of Creative Preservation?

As I introduced it at the beginning of this chapter, creative preservation refers to using creativity to ensure continuity in the affordances of relationships that make places habitable (Verger, Duymedjian, et al., 2024, 2025). It is a reflexive process insofar as people actively work to sustain those relationships that matters to them precisely because they perceive these relationships contribute to habitability. Creativity and preservation appear mutually constitutive: reflexive creative action sustains what is preserved (Verger, Duymedjian, et al., 2024; 2025), and what is reflexively preserved provides the conditions for further creativity (Le Masson et al., 2026). Those reflexive, creative practices involve preserving the affordances of relations by exploring new variations, arrangements, and connections with other relationships.

Initially, the creative preservation framework proposed by Verger, Duymedjian, et al. (2024) focused on *material and ideative continuity*, illustrated through practices such as upcycling, bricolage, low-tech approaches, and certain forms of craft for social justice. However, continuity in matter and ideas is always embedded within social, cultural, normative, institutional, and ecological domains, each with its specific affordances (Verger, Duymedjian, et al., 2025). We do not preserve things in themselves, but the social, cultural, and ecological possibilities they afford through these relationships. The creative preservation framework claims that, through a reflexive stance, we preserve various affordances (i.e., relationships that condition the ability to act, live and inhabit) that matter for people and the societies they form. What remains largely unexplored is the extent to which these affordances are preserved as the very material from which *societal futures* are constructed. This question is the pivot of the present section.

From the corpus of the present review, social–(ecological) innovation implies, indeed, a collective reflection about things to be changed or preserved. Venugopal and Viswanathan (2019), for instance, argue that social enterprises collaborate with local communities to 'endogenously determine aspects of local institutions that must be altered or preserved' (ibid., p. 802). For them, new cultural values and structures need not wholly displace existing ones, but may coexist alongside them, consistent with accounts of social change in which the old and the new persist 'without conflict and even with mutual adaptation' (ibid., p. 819). From my reading, this practice, as it seeks to reflect on what to preserve and what not to, is explicitly selective and normatively grounded. It does not seek to maintain relational continuity as such, but rather to retain those elements deemed socially desirable, while actively dismantling others. This is evident in the 'disrupting' stage identified by the authors, which involves the deliberate undermining and deinstitutionalisation of practices considered harmful, such as child marriage or restrictive health norms. In such instances, the intention is not to prevent decay, but to precipitate it within particular traditions, thereby clearing space for alternative arrangements. Its preservative impulse remains subordinated to normative judgements about which institutions warrant survival and which ought to be dismantled.

What I wish to foreground, here, is that these normative judgements are invariably future-oriented. The decision to dismantle child marriage or restrictive health norms secures the conditions under which different societal futures become possible. I am tempted to propose that creative preservation, when enacted at the scale of social–(ecological) innovation, operates as a form of ***anticipatory stewardship***, that is, a reflexive, collective process through which communities selectively sustain and rework the affordances of their socio-ecological milieu in order to keep viable (or collectively desirable and liveable) futures open. This anticipatory quality for habitability has not been sufficiently theorised in the creative preservation framework, yet it is precisely what gives social–(ecological) innovation its distinctive character as a scaled form of creative preservation. Societies do not simply preserve what they value because they also aim to sustain the futures they envision as worth inhabiting. To that end, people preserve the affordances that may lead them to live sustainably in those imagined futures.

This anticipatory dimension for habitability is visible in the work of Mongelli and Rullani (2017), who discussed the Italian project *Made in Carcere*, which collects waste materials from the fashion industry and employs convicted women to upcycle them into new fashion products sold to NGOs, institutions, and large companies. According to Mongelli et al. (2016), those who participated in the *Made in Carcere* project had a recidivism rate of 4%, compared with the Italian national average of 69% among former prisoners. This social innovation project appears to uphold the dignity and worth of convicted women by exploring the potential of waste materials to generate new economic value—that is, by creating opportunities for these upcycled items to be sold as a source of income. The *Made in Carcere* initiative provides an example of how social innovation opens a trajectory of future possibilities (reintegration, autonomy, and sustained livelihood for the person) that would otherwise be foreclosed.

We find similar anticipatory dynamics for habitability in Bosisio et al. (2024), who show that applying food-sharing preserves others' capacity to live meaningfully by sharing food rather than wasting it, which in turn has positive consequences for individuals' perceptions of their own worth and, I would add, for the collective capacity to imagine living in a desired future not governed by waste and scarcity.

A second contribution of the present reading of the socio-(eocological) innovation literature concerns the *intergenerational character* of creative preservation as a mechanism for addressing societal futures. The cases reviewed in this corpus point to the transmission across generations of the very relational affordances that enable future creativity (see also Lorenzi, Chap. 4, this Volume). Villalba Morales et al. (2023) showcase an initiative aimed at the social inclusion of marginalised communities. In proposing what they call 'inclusive innovation systems', they draw on the case of a group of peasants from a rural village in Colombia to show that peasants contribute traditional knowledge that can benefit conventional innovation systems. For Villalba Morales and colleagues, innovation systems can be inclusive when they incorporate agents who advocate for and actively support both the preservation of traditional knowledge and the appropriation of technological knowledge. What is traditional knowledge, if not the long accumulation of cultural heritage passed down through generations that structures the inventions people have made over the years? This transmission implies that each generation must reinterpret and re-express inherited affordances in light of new conditions. They preserve these inherited affordances because these affordances have ensured habitability for their elders, but they must reconfigure these same affordances because the network of relations that condition habitability has changed. New generations thus need to adapt this heritage to ensure its continuity. Vassallo et al. (2023), on their part, propose 'innovating to conserve' as a form of innovation that draws upon traditional knowledge in order to prevent the loss of biodiversity and cultural knowledge (ibid., p. 614). The expressions of new generations are layered over the preserved traces of what came before them. What the young generation makes of the affordances they inherit shapes the building blocks of the societal futures people eventually enact.

A third contribution of this review concerns the *affective and emotional infrastructure* that sustains creative preservation across societal scales. The creative preservation framework, as it stands, has emphasised reflexivity and the exploration of affordances that contribte to habitability (Verger, Duymedjian, et al., 2024). From the corpus reviewed here, it is clear that what motivates communities to engage in large-scale preservation is not merely a rational assessment of what is valuable, but a deeply felt concern (indeed, a form of care) about what might be lost and what might yet become possible. In this sense, creative preservation is tied to the notion of attachments in its broader sense, beyond humans. Lozano and Monsonís-Payá (2020) argue that we must act to secure the permanence of an authentic human life on Earth in the face of technological threats. Their work frames preservation as safeguarding the relationship humans have with their own existence and the socioecological milieu they inhabit, thereby preventing species-level 'decay' caused by unchecked innovation.

This form of care (what I would tentatively call *anticipatory concern*) is the emotional counterpart of the reflexive stance that creative preservation demands. Verger, Duymedjian, et al. (2024) briefly discussed this aspect of care in their article. Indeed, without this affective dimension, the notion of creative preservation risks becoming a purely cognitive, calculative exercise. The corpus reviewed in this chapter shows that communities preserve affordances not because they have calculated the utility of doing so, but because they care about what their affordances may become and how they may affect their lives. They care about preserving what their attachments afford in terms of habitability.

Roysen et al. (2024) provide further evidence of this affective underpinning. Their examination of how ecovillages employ creative practices to prevent ecological and material decay within local environments shows that an ecovillage in Ghana introduced clean cooking stoves and distributed tree seedlings to local communities to reduce the overharvesting of mangroves. This intervention aimed to maintain the ecological relationship between the community and the forest by reducing pressure on existing resources rather than replacing them. The motivation here seems to be a felt connection between people and their ecological milieu that drives creative action towards futures of sustained cohabitation.

Nakagawa and Kosaka (2022) demonstrated that crowdfunding platforms actively enable creative preservation by mobilising support for projects centred on heritage and regional revitalisation. Their analysis identifies recurring terms such as refurbishment, architecture, craftsman, and traditional, showing that project initiators deliberately direct creative activity towards maintaining and adapting existing cultural practices and built environments. In this configuration, creativity does not replace what exists but reworks and sustains it. The crowdfunding platform becomes a site that affords people to explore *which futures are worth preserving*. Sauermann et al. (2020) present several instances that align with this reading, particularly through repair and the maintenance of material and cultural continuity (an initiative called 'Re-paraKultur', which organised repair cafés to help citizens fix broken household items). These activities used creative, low-tech practices to prevent material decay while also encouraging reflection on everyday consumption. We find similar cases of collective action that shift norms from disposal towards reuse among Steinfield and Holt (2019). In each of these cases, the creative act entails preserving a shared image of what a sustainable, worthwhile future might look like. Again, creative preservation aims to preserve affordances, not things in themselves.

A fourth contribution that this corpus invites us to explore, and which I consider the most consequential for possibility studies, concerns the *political negotiation of societal futures* through creative preservation. If creative preservation is the process by which communities sustain the affordances from which their futures will be composed, then the question of *who* participates in this process, and on *whose terms*, becomes inescapable. Huguenin and Jeannerat (2017) showed that sustainability projects function as socio-technical and political sites where the purposes of sustainability are actively negotiated. Their work adds evidence that social-(ecological) innovation for sustainability can open up discussions about values, and Gallagher (2018) likewise show how, within such projects, actors debate what should count as valuable and work to make these values socially and economically meaningful. From the vantage point of creative preservation, these negotiations are fundamentally about selecting which relational affordances will be carried forward into the future and which will be allowed to decay as they don't serve habitability. This selective process is, I think, where the politics of creative preservation become most visible. Different actors bring different imaginaries of desirable futures, and the affordances they seek to preserve reflect these competing visions. Creative preservation, at scale, is politically laden. It is always a contested terrain where societal futures are at stake.

In some accounts, preservation is acknowledged yet subordinated to productivity. Vassallo et al. (2023), for example, frame some local knowledge about cultural burning as a practice that 'improves the health and productivity of the land' and has been 'passed from generation to generation… spawning a new industry'. Some initiatives appear socially innovative but remain embedded in 'business-as-usual' approaches that, for instance, open up 'biological populations and ecosystems for human usage in novel ways' (Bruynseels, 2020, p. 342). This example shows that preservation may contribute to creativity, as Le Masson et al. (2026) propose, but presumably for aims that place the biotope and biocenosis at the service of human activities, rather than for sustainable cohabitation (for a discussion of this theme, see Verger, Duymedjian, et al., 2025). Such approaches render natural 'resources' as substrates for innovation, thus echoing what this chapter identifies as a conservation-oriented approach to resources that serves innovation and profitability. When the societal future that creative preservation is oriented towards is one of continued extraction, the framework reveals its own limits: preservation is mobilised, but for a future that may undermine the very affordances for habitability it claims to sustain. This tension (between creative preservation as a genuine orientation towards meaningful societal futures and its co-optation by production logics) warrants sustained analytical attention in future research.

These examples show that preservation is not a static process, despite the term's common usage. Rather, preservation is enacted through a collective creative endeavour that interrogates what people hold dear (their attachments) and how these attachments contribute to a meaningful life and to habitability. From this premise, it is clear that social–(ecological) innovation expresses the value systems of societies (Sagiv & Schwartz, 2022; Verger, Duymedjian, et al., 2025). What the present analysis shows is that these value systems are *both retrospective evaluations of what matters and prospective orientations that shape the futures a society considers viable, desirable, and worth working towards to ensure habitability.* Creative preservation, when examined through the lens of social–(ecological) innovation, emerges as a fundamentally future-directed practice. The framework of creative preservation allows the study of what sustains the relational, material, and affective conditions from which meaningful, collective, and sustainable futures can be composed.

Creative Preservation and Societal Futures: Four Contributions

This chapter has shown that social–(ecological) innovation occupies a marginal position within mainstream innovation research, which remains largely oriented towards productivity, optimisation, and economic growth. A corpus analysis of 249 articles across leading innovation journals shows that social and socio-ecological concerns account for less than 5% of the sampled literature. Moreover, the language of conservation predominates over that of preservation and is primarily associated with the management of external resources. Preservation, by contrast, is more often linked to organisational continuity, identity, and internal structures. The chapter examined when social–(ecological) innovation may qualify as a scaled form of creative preservation and, crucially, proposed that creative preservation may be understood as a practice oriented not only towards the continuity of what is valued in the present, but towards the construction of meaningful societal futures.

The chapter makes four contributions to creative preservation research and to the broader field of possibility studies.

First, it proposes that creative preservation, when enacted at the scale of social–(ecological) innovation, is *inherently future-directed for habitability*. Societies both preserve the affordances of what they have and the futures they imagine as worth inhabiting. This anticipatory quality distinguishes creative preservation from mere conservation (a term which, as the corpus analysis has shown, tends to function as the management of external resources within existing value regimes).

Second, the chapter foregrounds the *intergenerational dimension* of creative preservation (a dimension that has remained largely unexplored in the framework). The cases reviewed here suggest that societal futures are built not on blank canvases but on inherited relational affordances, creatively reinterpreted by each generation as these affordances are important for making places habitable. The transmission and reworking of traditional knowledge, cultural heritage, and ecological practices across generations is itself a creative act and one that is essential to the construction of those viable futures. It is important, of course, to study how these reinterpretations do not completely erase the affordances of this heritage.

Third, the chapter identifies the *affective and emotional dimension (attachments)* that sustains creative preservation at societal scales. The emotional counterpart to the reflexive attitude that creative preservation requires is anticipatory concern for habitability, which is a felt orientation towards what might be lost and the ideas one might develop to preserve the affordances of what we care for. The cases of ecovillages, food-sharing initiatives, and repair cafés examined in this chapter show that communities endure because they care about the futures of people, materials, and ecosystems with which they are entangled.

Fourth, the chapter opens the question of the *political negotiation of societal futures* through creative preservation. If creative preservation is the process by which communities sustain the affordances that will compose their futures to ensure habitability, then it becomes central to addressing who participates in this process and on whose terms. The analysis suggests that creative preservation, at scale, is politically laden. It is always contested terrain where different imaginaries of desirable futures compete for the affordances to be carried forward. In some cases, preservation is subordinated to a 'meritocracy of production' (Verger, Duymedjian, et al., 2025), in which traditional knowledge and ecological practices are retained only insofar as they enhance efficiency or productivity. In others, innovation actively sustains socio-ecological relations through repair, adaptation, bricolage, and the maintenance of dignity and meaningful living. The reviewed body of work presents evidence that societies preserve various affordances through social–(ecological) innovation initiatives, but the *direction* of this preservation (the kind of future it serves) is subject to controversy. This chapter contributes to opening innovation research to considerations of preservation that are meaningfully accounted for by the creative preservation framework (Verger, Duymedjian, et al., 2024, 2025).

I am not reducing the question of creative preservation to a relativist account of value and judgement, namely, that everything would be worth preserving just because one person or a group of people thinks so. What some people may express as worth preserving may be the very thing that destroys other people's possibility to live meaningfully, but ultimately, the analyst should focus on how these actions modify the conditions for habitability for all species, not just humans. The project of creative preservation should aim to address the universality of how living beings preserve their means of sustaining the relations (and the affordances implied by these relations) that condition the habitability of social-ecological systems. This dynamic is evident in romantic relationships, parent–child bonds, and teacher–student support, where people actively work to maintain and renew the conditions that allow these relations to endure (Núñez-Regueiro et al., 2025; Verger, Roberts, et al., 2024, 2025; Verger & Duymedjian,

2020). We need more research that studies these effects beyond human systems.

I think there is something fundamentally true about the interplay between creation and preservation in how societies relate to one another and to their ecological milieu. Of course, people construct value systems around practices and seek to preserve what they hold dear. The question, however, is why this occurs, how it operates, when it emerges, and what it may inform scientific understanding about the fundamental dynamics of biological systems (within which humans, as other animals, are evidently included). Tentatively, I suggest that these questions are meaningfully addressed if we focus on how creative (and innovative) endeavours alter our shared conditions for habitability.

This chapter has proposed four perspectives as four contributions to both the project of creative preservation and the field of possibility studies. First, creative preservation, when understood as anticipatory stewardship for habitability, suggests that societies preserve the affordances that may lead them to the futures they imagine as worth inhabiting. Second, the intergenerational dimension of creative preservation emphasises that viable futures are built on the affordances provided by older generations for their habitability value, and questions whether/how/why the younger generation radically disrupts them or seeks to ensure continuity with these affordances through exploring novel arrangements suited to their contemporary experiences of what makes social-ecological systems habitable. Third, the affective dimension that sustains creative preservation at societal scales operates through anticipatory concern, namely, a felt orientation towards those attachements that might be lost and the idea one may come up with to ensure continuity in the affordances one holds dear. Fourth, the political negotiation of futures reveals that preservation is always a contested terrain, where competing imaginaries (from people with different levels of power) determine which affordances are to be preserved and which are not. These four contributions, I believe, provide essential coordinates for future research and take seriously the possibility that creative preservation is a fundamental feature of how living systems engage in creative practices to preserve the affordances that affect conditions for habitability.

References

Anthony, L. (2005). AntConc: Design and development of a freeware corpus analysis toolkit for the technical writing classroom. *IPCC 2005. Proceedings. International Professional Communication Conference, 2005*, 729–737. https://doi.org/10.1109/IPCC.2005.1494244

Bosisio, J., Zamparo, G., Mazzucchelli, A., Chierici, R., & Mason, M. C. (2024). Beyond economic convenience: Unveiling the motives for engaging in food sharing initiatives. *Technovation, 138*, 103119. https://doi.org/10.1016/j.technovation.2024.103119

Bruynseels, K. (2020). When nature goes digital: routes for responsible innovation. *Journal of Responsible Innovation, 7*(3), 342–360.

Dennis, M., Armitage, R. P., & James, P. (2016). Social-ecological innovation: Adaptive responses to urban environmental conditions. *Urban Ecosystems, 19*(3), 1063–1082. https://doi.org/10.1007/s11252-016-0551-3

Dennis, M., & James, P. (2018). Urban social-ecological innovation: Implications for adaptive natural resource management. *Ecological Economics, 150*, 153–164. https://doi.org/10.1016/j.ecolecon.2018.04.005

Edgeman, R., & Eskildsen, J. (2014). Social-ecological innovation. In *Encyclopedia of business analytics and optimization* (pp. 2231–2241). IGI Global Scientific Publishing. https://doi.org/10.4018/978-1-4666-5202-6.ch201

Gallagher, K. (2018). A reconsideration of social innovation: Drama pedagogies and youth perspectives on creative and social relations in Canadian schooling. *Canadian Journal of Education/Revue canadienne de l'éducation, 41*(1), 1–23.

Glăveanu, V. P. (2012). What can be done with an egg? Creativity, material objects, and the theory of affordances. *The Journal of Creative Behavior, 46*(3), 192–208. https://doi.org/10.1002/jocb.13

Huguenin, A., & Jeannerat, H. (2017). Creating change through pilot and demonstration projects: Towards a valuation policy approach. *Research Policy, 46*(3), 624–635. https://doi.org/10.1016/j.respol.2017.01.008

Kuzma, J., Gould, F., Brown, Z., Collins, J., Delborne, J., Frow, E., Esvelt, K., Guston, D., Leitschuh, C., Oye, K., & Stauffer, S. (2018). A roadmap for gene drives: Using institutional analysis and development to frame research needs and governance in a systems context. *Journal of Responsible Innovation, 5*(sup1), S13–S39. https://doi.org/10.1080/23299460.2017.1410344

Le Masson, P., Carvajal-Perez, D., Harlé, H., Hatchuel, A., Jibet, N., Taupin, L., & Weil, B. (2026). Design creativity for transitions: C-K/Topos, an advanced design theory for creative preservation. *Design Science, 12*, e6. https://doi.org/10.1017/dsj.2025.10045

Lozano, F., & Monsonís-Payá, I. (2020). Civic ethics as a normative framework for responsible research and innovation. *Journal of Responsible Innovation, 7*(3), 490–506. https://doi.org/10.1080/23299460.2020.1816024

Luria, S. R., & Kaufman, J. C. (2025). Social change creativity: Proposing a new domain for equity, social justice, and human rights. *The Journal of Creative Behavior, 59*(3), e70042. https://doi.org/10.1002/jocb.70042

Matthes, E. H. (2025). *What to save and why: Identity, authenticity, and the ethics of conservation*. Oxford University Press.

Melnykovych, M., Nijnik, M., Soloviy, I., Nijnik, A., Sarkki, S., & Bihun, Y. (2018). Social-ecological innovation in remote mountain areas: Adaptive responses of forest-dependent communities to the challenges of a changing world. *Science of the Total Environment, 613–614*, 894–906. https://doi.org/10.1016/j.scitotenv.2017.07.065

Mongelli, L., & Rullani, F. (2017). Inequality and marginalisation: Social innovation, social entrepreneurship and business model innovation: The common thread of the DRUID Summer Conference 2015. *Industry and Innovation, 24*(5), 446–467. https://doi.org/10.1080/13662716.2017.1295365

Mongelli, L., Versari, P., Rullani, F., & Vaccaro, A. (2016). Social Entrepreneurship and empowerment: Safe space and bridging in the case of madeincarcere. *Academy of Management Proceedings, 2016*(1), 14219. https://doi.org/10.5465/ambpp.2016.14219abstract

Mulgan, G. (2006). The process of social innovation. *Innovations: Technology, Governance, Globalization, 1*(2), 145–162. https://doi.org/10.1162/itgg.2006.1.2.145

Mulgan, G. (2012). The theoretical foundations of social innovation. In A. Nicholls & A. Murdock (Éds.), *Social innovation: Blurring boundaries to reconfigure markets* (pp. 33–65). Palgrave Macmillan. https://doi.org/10.1057/9780230367098_2.

Nakagawa, K., & Kosaka, G. (2022). What social issues do people invest in? An examination based on the empathy–altruism hypothesis of prosocial crowdfunding platforms. *Technovation, 114*, 102508. https://doi.org/10.1016/j.technovation.2022.102508

Núñez-Regueiro, F., Verger, N. B., Bressoux, P., Batruch, A., Bouet, M., Bressan, M., Brown, G., Butera, F., Cherbonnier, A., Darnon, C., Demolliens, M., de Place, A.-L., Desrichard, O., Goron, L., Hémon, B., Huguet, P., Jamet, E., Mazenod, V., Mella-Barraco, N., et al. (2025). How teacher autonomy support and student creativity jointly contribute to self-regulated learning: A dynamic, person-environment fit perspective. *Thinking Skills and Creativity, 58*, 101873. https://doi.org/10.1016/j.tsc.2025.101873

Olsson, P., & Galaz, V. (2012). Social-ecological innovation and transformation. In A. Nicholls & A. Murdock (Eds.), *Social innovation: Blurring boundaries to reconfigure markets* (pp. 223–247). Palgrave Macmillan UK. https://doi.org/10.1057/9780230367098_10

Phillips, W., Lee, H., Ghobadian, A., O'Regan, N., & Et., A. (2014). Social innovation and social entrepreneurship. *Group & Organization Management*. https://doi.org/10.1177/1059601114560063

Roysen, R., Bruehwiler, N., Kos, L., Boyer, R., & Koehrsen, J. (2024). Rethinking the diffusion of grassroots innovations: An embedding framework. *Technological Forecasting and Social Change, 200*, 123156. https://doi.org/10.1016/j.techfore.2023.123156

Sagiv, L., & Schwartz, S. H. (2022). Personal values across cultures. *Annual Review of Psychology, 73*, 517–546. https://doi.org/10.1146/annurev-psych-020821-125100

Sauermann, H., Vohland, K., Antoniou, V., Balázs, B., Göbel, C., Karatzas, K., Mooney, P., Perelló, J., Ponti, M., Samson, R., & Winter, S. (2020). Citizen science and sustainability transitions. *Research Policy, 49*(5), 103978. https://doi.org/10.1016/j.respol.2020.103978

Schot, J., & Steinmueller, W. E. (2018). Three frames for innovation policy: R&D, systems of innovation and transformative change. *Research Policy, 47*(9), 1554–1567. https://doi.org/10.1016/j.respol.2018.08.011

Schumpeter, J. A. (1962). *Capitalism, socialism, and democracy* (3rd ed.). Harper Perennial.

Steinfield, L., & Holt, D. (2019). Toward a theory on the reproduction of social innovations in subsistence marketplaces. *Journal of Product Innovation Management, 36*(6), 764–799. https://doi.org/10.1111/jpim.12510

Tidd, J., & Bessant, J. R. (2021). *Managing innovation: Integrating technological, market and organizational change* (7th ed.). Wiley.

Vassallo, J. P., Banerjee, S., & Prabhu, J. C. (2023). Biocultural innovation: Innovating at the intersection of the biosphere and ethnosphere. *Journal of Product Innovation Management, 40*(5), 610–629. https://doi.org/10.1111/jpim.12669

Venugopal, S., & Viswanathan, M. (2019). Implementation of social innovations in subsistence marketplaces: A facilitated institutional change process model. *Journal of Product Innovation Management, 36*(6), 800–823. https://doi.org/10.1111/jpim.12508

Verger, N. B. (2024). Transformational creativity, care, and the common good: Toward a refined definition? *The Journal of Creative Behavior*, jocb.1511. https://doi.org/10.1002/jocb.1511

Verger, N. B., & Duymedjian, R. (2020). A theoretical framework of romantic creativity: Dyadic creativity in romantic relationships and plausible links with wellbeing. *International Journal of Wellbeing, 10*(5). https://doi.org/10.5502/ijw.v10i5.1493

Verger, N. B., Duymedjian, R., & Glaveanu, Vlad P. (2025). The meritocracy of preservation: Reimagining merit beyond production. *Organization*. Advanced online publication. https://doi.org/10.1177/13505084251347519

Verger, N. B., Duymedjian, R., Wegener, C., & Glăveanu, V. P. (2024). Creative preservation: A framework of creativity in support of degrowth. *Review of General Psychology, 28*(3), 268–281. https://doi.org/10.1177/10892680241256312

Verger, N. B., Roberts, J., Guiller, J., & McAloney-Kocaman, K. (2024). The whys and wherefores of home creativity support: A cross-cultural reflexive thematic analysis between British and French parents. *Thinking Skills and Creativity, 52*, 101527. https://doi.org/10.1016/j.tsc.2024.101527

Verger, N. B., Roberts, J., Guiller, J., & McAloney-Kocaman, K. (2025). Parent–child creative activities are associated with children's positive affect and relationships quality. *Psychology of Aesthetics, Creativity, and the Arts.* https://doi.org/10.1037/aca0000790

Villalba Morales, M. L., Ruiz Castañeda, W., & Robledo Velásquez, J. (2023). Configuration of inclusive innovation systems: Function, agents and capabilities. *Research Policy, 52*(7), 104796. https://doi.org/10.1016/j.respol.2023.104796

Wegener, C. (2023). Upcycling. In V. P. Glăveanu, L. Tanggaard, & C. Wegener (Éds.), *Creativity—A new vocabulary* (pp. 273–282). Springer International Publishing. https://doi.org/10.1007/978-3-031-41907-2_25.

CHAPTER 3

Crealectics: A Method for Compossible Worlds

Luis de Miranda

Introduction

The boundary between what is and what could be shimmers with potential. For centuries, humanity has enacted and celebrated its capacity to make the *unimpossible* possible. We have become masters of possibility, architects of the previously unthinkable. Yet as we survey the horizon of our achievements, a troubling pattern emerges: our local marvels have sometimes evolved into global catastrophes or threats. Our innovations can become our predicaments. I believe the challenge of our time is not merely to continue generating innovations but to cultivate what the seventeenth-century philosopher Gottfried Wilhelm Leibniz (1714/1989) termed 'compossibility': the capacity to create diverse goods, values, and practices that can coexist in mutually enhancing rather than mutually destructive or contradictory relationships.

The smartphone, for instance, connects us to an infinite supply of information while simultaneously isolating us from embodied and shared

L. de Miranda (✉)
Centre for Research Ethics and Bioethics, Uppsala University, Uppsala, Sweden
e-mail: luis.demiranda@uu.se

N. B. Verger, D. Mattson (eds.), *Possibility Studies and Societal Futures*, Palgrave Advances in Possibility Studies,
https://doi.org/10.1007/978-3-032-26379-7_3

presence (Turkle, 2011). Over time, this reveals an incompossibility between digital connectivity and physical togetherness. Industrial systems lifted millions out of poverty yet now threaten the stability of Earth's climate, with global material prosperity perhaps incompossible with atmospheric stability (Steffen et al., 2015). Financial instruments democratised investment while creating unprecedented inequality (Piketty, 2014): access to markets remains to be shown to be compossible with equitable distribution. Social media platforms promised to unite humanity but have fractured our sense of reality and togetherness (Tufekci, 2017)—connection networks seem more compossible with tribalism than with global epistemic coherence or a planetary common sense.

It can, of course, be argued that these same innovations also demonstrate potential compossibility: smartphones can facilitate rather than replace presence when used responsibly; industrial capacity can serve ecological regeneration when redesigned; financial systems may distribute rather than concentrate wealth through different governance structures. The question is whether we can create or engineer conditions for their compossibility. Each breakthrough, developed and considered in isolation, represents a triumph of human ingenuity. But the challenge remains: viewed together, technological possibilities often reveal what we may call a "compossibility crisis," a challenge to ensure that our diverse innovations can coexist healthily and justly within the same world.

This crisis extends far beyond technology into every domain of human endeavour, affecting not only material goods but also values, practices, identities, narratives, and ways of knowing. Political movements that individually champion worthy causes such as economic justice, environmental protection, cultural preservation, and individual freedom often find themselves in irreconcilable conflict when their various agendas meet in the arena of policy and practice (Mouffe, 2000). Academic disciplines have generated profound insights through specialised focus yet struggle to integrate their knowledge into coherent wisdom to address complex, interdisciplinary challenges (Klein, 1990). Every day, organisations that excel at optimising specific metrics (efficiency, growth, innovation, sustainability) find that local optimisation can lead to global dysfunction. What Verger et al. (2025) identify as the tension between the 'meritocracy of production' (which privileges growth, novelty, and accumulation) and the 'meritocracy of preservation' (which values maintenance, care, and continuity) exemplifies the structural incompossibility that operates across economic, social, and cultural systems.

Why has compossibility become critically urgent now? Three interrelated factors distinguish our contemporary situation from previous eras. First, the pace of innovation has accelerated beyond our collective capacity to assess systemic implications. In earlier centuries, technologies evolved gradually, allowing incompossibilities to reveal themselves over generations and sometimes enabling organic cultural adaptation and institutional adjustment. Now we generate and globally deploy innovations (from gene editing to artificial intelligence) faster than we can understand their interactions. Second, the scale of our interventions has become planetary. Local innovations in agriculture, energy, manufacturing, or communication now propagate through interconnected global systems, turning what were once geographically contained problems into cascading crises that recognise no boundaries (Steffen et al., 2015). A financial innovation in New York triggers food riots in Cairo; an industrial process in one hemisphere alters atmospheric chemistry, affecting weather patterns everywhere. Third, our systems have achieved what Perrow (1999) terms 'tight coupling', in which technological, economic, and social elements have become so interdependent that disturbances in one domain rapidly cascade throughout entire networks before we can intervene. These three factors (acceleration, globalisation, and tight coupling) mean that the mindset enabling our species' remarkable advances, the capacity to imagine and realise what once seemed impossible, has become insufficient for navigating a world where everything affects everything else. We have mastered making things possible, but we have not yet learned to make them compossible, compatible with other valuable possibilities within shared lifeworlds.

This requires what Verger et al. (2024) call 'creative preservation', approaches that honour both innovation and continuity, and transformation and maintenance, recognising that genuine progress often involves sustaining valuable practices while transforming unsustainable ones (see also Verger, Chap. 2, this volume). Creativity need not serve endless accumulation but can instead be directed towards meaningfully regenerating what sustains life. The crealectic method provides a practical pathway for enacting such practice, offering processual phases to navigate the productive tension between transformation and care, and between creating new possibilities and sustaining existing balances. Crealectics is a systematic approach to fostering compossible thinking and action. Grounded in process philosophy (Whitehead, 1978; Bergson, 2007), crealectics offers a five-phase methodology for moving beyond either-or thinking towards both-and synergies and beyond competitive fragmentation towards collaborative intercreation. The method has been pilot-tested in organisational contexts since 2023, ranging from an energy company navigating the transition to renewable energy systems (our case study in this chapter) to educational and healthcare institutions.

This chapter comprises three main sections. First, I establish theoretical foundations by tracing the concept of compossibility from Leibniz through process philosophy to contemporary systems thinking. Second, I introduce the crealectic method's five phases (resetting, crealing, profusing, compossibilising, and realising), explaining how each addresses distinct aspects of the innovation challenge while contributing to a holistic, iterative process. Finally, I reflect on broader implications for enacting societal change, emphasising how this approach differs fundamentally from both analytical problem-solving (which dissects systems into isolated components for separate optimisation) and dialectical synthesis (which resolves contradictions primarily through oppositional tension). Throughout, I show how crealectics provides a methodological framework for what anthropologist Arturo Escobar (2018) calls 'designs for the pluriverse'—approaches that enhance rather than diminish the multiplicity of ways life can flourish.

But why 'crealectics' rather than simply a 'compossible method'? This terminological choice reflects a crucial conceptual distinction: compossibility denotes the desired state in which multiple valuable innovations, practices, and ways of being can coexist eudynamically without mutual destruction or subordination. Crealectics—grounded in the axiom of 'Creal', the Creative Real—names the process. Where analytics dissects reality into manageable parts for separate analysis and optimisation, and where dialectics synthesises opposites through conflict, negation, and resolution, crealectics weaves diverse elements into living wholes that preserve and enhance their distinctive qualities through not only interaction but also intercreation.

At its core, the crealectic method addresses what may be the defining challenge of enacting societal change in the twenty-first century: how to preserve what we value while transforming what we must; how to honour diversity while creating coherence; how to foster innovation while ensuring holistic compatibility. The meritocracy of preservation, as conceptualised by Verger et al. (2025), recognises the value of maintaining infrastructure, caring for ecosystems, and sustaining communities, work that the dominant meritocracy of production often renders invisible or devalues. Crealectics offers both a theoretical framework and practical tools for navigating and integrating these complementary logics.

Changing the world begins with transforming our relationship with change itself. The quality of attention, intention, and relational capacity we bring to innovation processes fundamentally shapes the outcomes we generate (Scharmer, 2009). Before we can create genuinely new systems, we must often examine and shift the mental models, emotional patterns, perceptual habits, and relational dynamics that produce incompossibility in the first place. Systemic transformation requires not only new techniques but also a new collective consciousness.

Theoretical Foundations: From the Possible to the Compossible

Leibniz's Compossibility and Modern Challenges

The concept of compossibility emerged from one of the most ambitious philosophical projects in Western thought: Gottfried Wilhelm Leibniz's attempt to reconcile divine perfection with worldly imperfection. Writing in the early eighteenth century, Leibniz grappled with a question that resonates powerfully in our current era of global interconnection: given the infinity of possible worlds, what principle guides the selection of our reality? His answer, compossibility, offers crucial insights into contemporary challenges of societal transformation.

For Leibniz (1989), compossibility was more than mere logical consistency. While two propositions might be individually possible (a world with only red objects, a world with only blue objects), they become compossible only when they can coexist within the same coherent reality without fundamental contradiction. Leibniz argued that our world represents 'the best of all possible worlds' not because it contains no suffering or imperfection, but because it achieves the greatest diversity of existence, with elements functioning harmoniously together and maximising actuality through compossible variety. This is not about optimisation in the narrow sense of maximising a single metric, but rather about what we might call 'symphonic maximisation': creating the richest possible composition in which multiple distinct elements enhance rather than diminish each other's existence. A symphony achieves its power not by making every instrument louder but by enabling each to contribute its distinctive voice to a coherent whole.

This principle proves prescient for understanding our contemporary predicament. The climate crisis, for instance, arises from the incompossibility of multiple individually beneficial innovations: industrial agriculture that feeds billions while depleting soil and water systems; transportation networks that enable global connectivity while destabilising atmospheric chemistry; energy systems that power modern civilisation while disrupting planetary climate patterns (Rockström et al., 2009). Leibniz's framework helps us recognise that the challenge is not to eliminate possibility but to cultivate discernment about which possibilities belong together. While we need not adopt Leibniz's theological commitments, we can appreciate the profound insight embedded in his vision: genuine eudynamia (etymologically from 'good' and 'potential') requires not just an abundance of goods but wisdom about their mutual compatibility.

Process Philosophy and Relational Ontology

Twentieth-century process philosophy offers further conceptual resources for understanding compossibility. Alfred North Whitehead (1978) reconceived reality as fundamentally composed not of static substances but of dynamic processes, 'occasions of experience' that achieve their identity through their relationships with other processes. In this view, things do not first exist independently and then enter into relations; rather, their very being emerges through and as relationship.

This relational ontology must have implications for our understanding of innovation and change. If entities realise their reality through their relationships, the introduction of new elements necessarily reconfigures the entire relational field. A new technology does not merely add to what already exists; it transforms the meaning and function of existing elements by altering their relational context. The smartphone, for example, did not merely provide a new communication tool; it reconfigured the human experience of attention, presence, intimacy, and public space (Turkle, 2011).

Henri Bergson's (2007) complementary vision of 'creative evolution' adds crucial temporal dimensions to process thought. Bergson argued that genuine creativity involves not only recombining existing elements but also generating novel forms that could not have been predicted from prior conditions. His concept of 'duration' (durée), designating time as lived rather than a quantitative succession, helps us understand why compossible innovation requires attention to lived experience rather than merely measurable outcomes. From the point of view of duration, all beings are part of a continuum of intercreation.

Complexity Theory and Systemic Interdependence

As far as I can tell, contemporary complexity theory offers empirical and mathematical support for the philosophical insights of process philosophy. Research in complex adaptive systems shows that in highly interconnected networks, local optimisations often produce system-level dysfunction, a phenomenon related to 'frustration' in physics, where competing constraints prevent optimal solutions (Kauffman, 1995). When each component of a system pursues its local optimum without regard for systemic effects, the result can be global configurations in which no component achieves its potential and the system as a whole fails to achieve eudynamia (healthy compossibility).

Stuart Kauffman's (1995) work on 'fitness landscapes' illustrates this dynamic. In systems with multiple interdependent variables, the fitness of any configuration depends not only on the intrinsic qualities of its components but also on how those components combine with others. Some combinations create 'fitness peaks', configurations in which multiple elements enhance each other's functioning. Other combinations create 'fitness valleys', configurations in which elements undermine each other. From a eudynamic perspective, the challenge of compossible innovation is to navigate from existing configurations towards fitness peaks where diverse elements enhance rather than diminish collective flourishing.

Network theory adds additional insights about systemic interdependence. Research demonstrates that in highly connected networks, effects propagate nonlinearly: small perturbations can trigger cascading changes, whereas large interventions sometimes produce negligible results (Barabási, 2002). This non-linearity means that assessing the compossibility of innovations requires understanding not just their direct effects but their potential to trigger cascading transformations throughout interconnected systems.

Finally, ecological systems thinking emphasises the importance of 'emergent properties' (Odum, 1969), characteristics of systems that arise from relationships between components rather than being reducible to those components. A forest ecosystem's resilience, for instance, emerges from the relationships among diverse species, not from the properties of individual trees. Compossible innovation must attend to how new elements affect such emergent properties, recognising that undermining systemic resilience or coherence constitutes a form of incompossibility even if no individual component fails.

Towards a Philosophy of Intercreation

These theoretical streams converge in a process philosophy of intercreation, a framework that recognises creation as fundamentally collaborative, relational, and participatory rather than individualistic, autonomous, and imposing. Process philosophy emphasises that entities do not interact as pre-existing, independent agents but co-constitute one another through their ongoing relationship. Agency is not a property possessed by individuals but emerges through relationships and is underpinned by a foundational creative flux.

Applied to innovation, this philosophy holds that genuinely compossible creation requires attending to how new possibilities co-evolve with existing realities rather than imposing predetermined designs. It recognises that we are always creating within systems that are simultaneously creating us, and that our innovations are responses to and participations in larger creative processes rather than sovereign acts of will imposing order on passive matter.

As a philosophy of intercreation, crealectics echoes Indigenous knowledge systems that have long recognised the fundamental interdependence of all forms of life (Kimmerer, 2013). Many ancestral cosmologies understand humans as participants rather than masters of creation, emphasising reciprocity, responsibility, and gratitude as foundational to sustainable innovation (Cajete, 2000). For instance, the principle of considering impacts on 'seven generations' reflects a sophisticated understanding of temporal compossibility, recognising that present innovations must be compatible with the flourishing of future generations. Generativity is not only the capacity to create but also the capacity to care for future creatures.

This diverse theoretical foundation suggests that compossibility is not merely a pragmatic consideration for improving system effectiveness but also reflects fundamental truths about the nature of reality. We inhabit a universe of relationships, processes, interdependence, and intercreation. Crealectics emerges as a tentative methodology to align human innovation with these deeper patterns of reality, enabling us to participate consciously in the larger creative processes through which life evolves, adapts, and flourishes.

The Crealectic Method: Five Phases of Compossible Innovation

The crealectic method (de Miranda, 2025) offers a structured yet flexible pathway from incompossible fragmentation to compossible integration. Its five phases (resetting, crealing, profusing, compossibilizing, and realizing) form an iterative spiral rather than a linear sequence. Each phase addresses distinct aspects of the innovation challenge while contributing to a holistic transformation that integrates personal development, collective creativity, and systemic change.

Phase 1: Resetting

The first phase, resetting, involves deliberately creating openness to new possibilities by releasing attachment to existing patterns, assumptions, and solutions that may have become obstacles to compossible innovation. This entails examining the 'interior condition' from which we approach challenges (Scharmer, 2009), including the unexamined assumptions, habitual thought patterns, and emotional investments that constrain our perception of what is possible.

Resetting draws on René Descartes' methodological doubt, the radical questioning that led to his famous cogito ergo sum, 'I think, therefore I am' (Descartes, 1637/1999). However, whereas Descartes sought absolute certainty through systematic doubt, resetting seeks suspension of belief. The goal is not to eradicate all beliefs but to cultivate what Zen Buddhism describes as the 'beginner's mind' (*shoshin*), a state of openness and receptivity unclouded by preconceptions (Suzuki, 1970).

In practice, resetting can involve various forms of systematic inquiry, such as:

Assumption auditing: Teams explicitly surface and examine the unstated premises that underpin their current approaches. What do we take for granted about how change happens? About what stakeholders value? Which resources or constraints are fixed, and which ones are variable? These assumptions, often invisible until deliberately examined, profoundly shape what innovations seem possible or impossible.

Beginner's mind cultivation: Drawing on the Zen Buddhist practice of shoshin (Suzuki, 1970), resetting cultivates what Shunryu Suzuki called the 'beginner's mind', approaching even familiar situations with fresh openness rather than the accumulated certainties that can blind us to new possibilities. This does not mean abandoning expertise but rather holding it lightly enough to perceive what it might exclude.

Shadow work: Following Carl Jung's (1968) insight that what we deny or repress often shapes our behaviour more powerfully than what we consciously acknowledge, resetting involves examining which aspects of a situation we have been avoiding, denying, or marginalising. What perspectives have we dismissed as unrealistic? What values have we deemed incompatible with progress? Often, these shadow elements contain crucial information about the incompossibilities we generate through selective attention.

The psychological challenge of resetting lies in letting go of certainty and control, which can generate anxiety and resistance. Neuroscience research on insight and creativity explains why this discomfort is necessary. Mark Jung-Beeman et al. (2004) found that creative insights involve temporarily suspending focused analytical thinking to allow diffuse associative processes to generate novel connections. Resetting creates the cognitive and emotional space for such insights by interrupting the momentum of established patterns.

A crucial aspect of resetting, as David Bohm (1996) called it, inspired by Husserlian phenomenology, is 'suspension': holding views in awareness with enough distance for alternative possibilities to emerge. This differs from dialectical negation, which seeks to overcome existing positions by opposing them. Resetting creates a clearing, a space of not-yet-knowing, from which genuinely new possibilities may or may not arise, rather than merely rearranging existing elements. Resetting is also a way of acknowledging that one does not need to be in a constant state of productivity.

The output of resetting is not a new solution but a new quality of awareness, a way of attending to situations in order to perceive previously invisible dimensions. This expanded awareness becomes the foundation for the next phase: crealing.

Phase 2: Crealing

The second phase, crealing (combining 'Creal'—Creative Real—and 'feeling'), involves reconnecting with the creative flow underlying reality itself. This phase deliberately emphasises both tapping into cosmic creativity and feeling the abundant potential that supports compossible innovation, often through emotions such as gratitude for being alive and the capacity to engage creatively with what is emerging.

Crealing operates on the principle that possibility becomes more accessible when we first reconnect with the universal source of creative potential rather than immediately attempting to generate specific solutions. By attuning to this fundamental creative abundance, we enhance our embodied capacity to recognise patterns of compatibility later. The phase draws on embodied cognition research demonstrating that felt experience and somatic awareness enhance creative capacity (Lakoff & Johnson, 1999; Varela et al., 1991). When participants reconnect with their bodily sense and the living flow of experience, they access deeper sources of becoming. The psychological mechanism involves what Gendlin (1978) called the 'felt sense', the capacity to perceive meaning through bodily awareness, a form of intuition that deepens when we temporarily suspend analytical pressure.

Practical methods for crealing include:

Evolutionary Visualisation: Guiding participants through a visualisation moving backwards down the tree of life, from humans to earlier life forms, and ultimately to the origin of life itself. Research by Barbara McClintock (1984) demonstrated that deep identification with living systems enhances insight. This practice helps people feel their connection to the continuous creative process shaping all life.

Natural Meditation: Drawing on Bachelard's (1958) work on the material imagination, groups engage directly with natural elements (water, wind, earth…). At the Älvkarleby dam with Vattenfall engineers, participants closed their eyes and listened to the rushing water as the 'voice of the Creal'. How do natural flows embody creative potential? These practices cultivate what Abram (1996) calls 'sensory participation', reconnecting with the more-than-human creative processes that inspire innovation.

Collective Randomness: Following techniques from Surrealist automatic writing (Breton, 1969), groups use random word combinations to bypass habitual thought patterns. Each participant selects words at random and assembles them into phrases related to the workshop theme, creating a 'multiverse poem'. This practice dissolves barriers between individual imaginations, creating a shared field of creative potential that reveals unexpected connections.

Biomimicry and Flow Recognition: Creating quick, low-fidelity explorations of how natural systems embody creative principles helps groups learn from nature's compossible patterns. Design thinking pioneer Janine Benyus (1997) showed how observing nature's solutions, for instance, ecosystems coordinating without central control and organisms balancing efficiency and resilience, can inspire innovations that work with rather than against natural processes, thereby enhancing compatibility with living systems.

Musical Immersion: Music directly accesses the pre-conceptual, flowing quality of experience that characterises the Creal. Drawing on Aldridge's (1996) research on music and consciousness, facilitators use carefully selected compositions to help participants bypass purely analytical thinking and connect with embodied creativity flow.

Phase 3: Profusing

The third phase, profusing, involves the uncensored generation of abundant possibilities without immediate judgment. This phase deliberately encourages free association and divergent thinking, allowing ideas to pour forth in cascades of creative expression within a given thematic frame. Profusing operates on the principle that compossibility emerges more readily from abundance than from scarcity. By generating diverse possibilities first, without premature filtering, participants create the rich material from which compossible patterns can later be discovered. Creative systems thrive through variation first, followed by selection (Kauffman, 1995; Sawyer, 2012).

The phase also draws on divergent-thinking research showing that quantity breeds quality in ideation (Guilford, 1967; Osborn, 1953). When groups generate many ideas without immediate evaluation, they uncover unexpected combinations and novel possibilities. The psychological mechanism is what Carson (2011) termed 'cognitive disinhibition', whereby mental filters are temporarily reduced, allowing unusual associations to enter consciousness.

Practical methods for profusing include:

Individual Written Profusion: Participants write as many ideas as possible related to the theme, without pausing to evaluate them. Research by Mueller and Oppenheimer (2014) suggests that writing by hand engages different neural pathways from typing, potentially enhancing creative thinking. The physical act of writing slows thought just enough to capture ideas more fully.

Round-Robin Sharing: In small groups, participants take turns sharing one idea at a time, building on each other's contributions. This structured approach ensures equal opportunity to contribute and prevents dominant personalities from monopolising the conversation. This method produces more creative outcomes than unstructured group brainstorming (Paulus and Nijstad, 2003).

Reverse Thinking: Participants deliberately generate 'wrong' answers or opposite approaches to the challenge. Following de Bono's (1970) work on lateral thinking, this technique deliberately disrupts conventional thinking patterns. By temporarily inverting assumptions, groups uncover novel perspectives that might not emerge through direct approaches.

Visual Profusing: Using drawing, mind mapping, or collage to express ideas visually rather than verbally. This approach engages spatial reasoning and pattern recognition, complementing verbal ideation. Root-Bernstein and Root-Bernstein (1999) demonstrated that visual thinking tools make the familiar strange and reveal hidden connections.

Phase 4: Compossibilising

The fourth phase, compossibilising, involves organising diverse possibilities into dynamically harmonious, coherent patterns. This phase deliberately seeks compatibility among different elements, creating eudynamic configurations across the dimensions of person, earth, work, and society.

Compossibilising operates on Leibniz's (1989) principle that coherent worlds comprise elements that can coexist without fundamental contradiction; they are 'possible together' or compossible. By identifying potential synergies among diverse ideas generated during profusion, this phase seeks patterns of compatibility and mutual enhancement, analogous to how complex adaptive systems generate emergent properties through interactions among elements (Holland, 2014; Kauffman, 2019). The phase draws on systems thinking research showing that wholes exhibit properties their parts cannot (Capra & Luisi, 2014; Meadows, 2008).

When participants examine relationships among possibilities rather than possibilities in isolation, they discover how seemingly disparate elements might form coherent systems. The psychological mechanism involves what Alexander (2002) called 'pattern language recognition', the perception of recurring structures that create living coherence. Alexander identified fifteen fundamental properties (including 'levels of scale', 'strong centres', 'good shape', and 'local symmetries') that consistently appear in spaces and systems that feel alive and coherent. These properties work together, not as rigid rules but as flexible principles that can be adapted across contexts. Pattern language recognition entails learning to perceive deep structural properties that make systems appear coherent and generative rather than fragmented or dead. In compossibilising, participants develop sensitivity to similar patterns of compatibility, recognising when diverse elements share underlying structural affinities that allow them to enhance rather than obstruct one another, creating configurations that possess what Alexander called 'the quality without a name', a difficult-to-articulate but unmistakable sense of rightness and vitality that professional composers cultivate.

Practical methods for compossibilising include:

Affinity Clustering: Participants individually identify connections between ideas, then collaboratively refine these clusters. This approach balances individual insight with collective intelligence, preventing domination by single perspectives while allowing expertise to inform the process. Research by Snowden and Boone (2007) showed that this method effectively reveals emergent patterns.

World's Creation: Small teams each develop a composable model, a coherent configuration of compatible possibilities, then share insights across teams. This technique, adapted from scenario planning (Schwartz, 1991), enables the simultaneous exploration of multiple potential futures. Groups discover which elements consistently appear across scenarios, suggesting robust compossibilities.

Systems Mapping: Using visual tools to map relationships among possibilities, revealing potential synergies and conflicts. Drawing on Senge's (1990) work on systems thinking, participants create diagrams that show how different elements might interact dynamically. This practice makes implicit assumptions about compatibility explicit for examination.

Eudynamic Assessment: Evaluating emerging patterns against the four dimensions of the CIPHER framework: person (embodied), earth (extended), work (enactive), and society (embedded). Following the 4E cognition perspective (Newen et al., 2018), participants assess whether configurations enhance balance across these dimensions and identify where adjustments might improve overall coherence.

Phase 5: Realising

The fifth phase, realising, involves initiating the implementation of compossible innovations and embedding them within organisational, social, and material systems to enable their continued flourishing. This is where vision becomes reality, possibility becomes actuality, and creativity becomes creation.

Realising is often treated as straightforward execution, simply applying sufficient resources and effort to implement predetermined plans. However, decades of implementation research reveal that the gap between designed solutions and realised outcomes reflects fundamental challenges rather than mere execution failures (Pressman & Wildavsky, 1984). Plans inevitably encounter unanticipated conditions; stakeholders interpret innovations differently from what is intended; and context transforms both innovations and implementers.

The crealectic approach to realising does recognise implementation as itself a creative, adaptive process requiring ongoing learning and adjustment. It draws on Donald Schön's (1983) concept of 'reflection-in-action', that is, the practitioner's capacity to observe and respond to emerging situations in real-time rather than rigidly following predetermined procedures.

Practical methods for realising include:

Prototyping and iteration: Instead of implementing innovations at full scale immediately, groups create small-scale prototypes that enable learning through experimentation (Brown, 2009). Each iteration reveals unexpected challenges and opportunities, allowing refinement before major resource commitments. This approach, common in design thinking, treats implementation as discovery rather than mere execution.

Adaptive management: Drawing on ecological management literature (Holling, 1978; Walters, 1986), groups treat implementation as an ongoing experiment, establishing monitoring systems that provide feedback on whether innovations are generating intended effects and acceptable impacts. When monitoring reveals unintended consequences or better opportunities, the approach adjusts, treating plans as hypotheses to be tested rather than blueprints to be followed.

Participatory implementation: Groups involve affected stakeholders not only in designing innovations but also in adapting them during implementation (Chambers, 1997). This builds on the insight that those closest to situations often perceive dimensions that distant designers miss.

Participatory approaches also foster stakeholder ownership and commitment, recognising that successful implementation requires not only technical adequacy but also social legitimacy.

Embedding in infrastructure: Groups work to institutionalise compossible innovations within organisational structures, policies, budgets, and physical infrastructure, so they become the new normal rather than requiring constant voluntary effort (Schein, 2010). This might involve changing hiring criteria, modifying evaluation systems, redesigning physical spaces, or reallocating resources. The goal is to make compossible approaches easier than incompossible ones, shifting default patterns rather than relying on exceptional individual commitment.

Narrative and meaning-making: Groups actively shape the stories told about innovations, recognising that the meanings people assign to changes profoundly affect their reception and sustainability (Bruner, 1990). Is this innovation framed as a sacrifice or an opportunity? As abandoning tradition or honouring deeper values? As technocratic imposition or democratic expression? Effective realizing involves sense-making, helping communities develop shared interpretations that enable coordinated action.

Cultivating practice communities: Groups establish ongoing forums where practitioners of new approaches can share insights, troubleshoot challenges, and develop collective expertise (Wenger, 1998). Compossible innovations often require new skills, habits, and forms of collaboration that cannot be fully specified in advance but must be developed through practice. Communities of practice provide the social infrastructure for such ongoing learning.

The temporal challenge of realising involves what organisational scholar James March (1991) called the tension between 'exploitation and exploration', that is, between refining existing approaches (exploitation) and discovering new possibilities (exploration). Realising necessarily emphasises production as innovations become established, but this must be balanced with continued exploration that allows adaptation and pruning as contexts change (see for instance, Verger et al., 2024). Systems that over-optimise for current conditions become brittle and unable to respond when conditions shift. Realising also involves attention to recognising that innovations become sustainable only when they enhance rather than deplete the systems within which they are embedded. This means assessing not only whether innovations work technically but also whether they regenerate the ecological, social, economic, and cultural resources on which they depend.

Finally, the spiritual dimension of realising involves what phenomenologist Martin Heidegger (1977) called 'releasement' (*Gelassenheit*), that is, a quality of presence that allows things to emerge according to their own nature rather than imposing predetermined forms. This paradoxical combination of intention and non-attachment enables groups to work purposefully towards goals while remaining responsive to what emerges. It reflects the Taoist principle of *wu wei*, effortless action that works with rather than against the grain of situations (Watts, 1975).

The output of realising is not just implemented innovations but transformed systems, organisations, communities, or sectors that now embody compossible approaches as their normal mode of operation. These realised changes, in turn, become the new context for subsequent cycles of creative innovation, as groups periodically reset, create, proliferate, compose, and realise in response to emerging challenges and opportunities.

Crealectic Consciousness and Societal Transformation

Having explored the crealectic method's theoretical foundations and practical phases, we can now reflect on the broader implications for enacting societal change in an era of systemic interdependence and ecological limits.

From Problem-Solving to Possibility-Creating

Conventional approaches to societal change typically frame challenges as problems to be solved, obstacles to be overcome through sufficient technical ingenuity, political will, or resource mobilisation. This problem-solving orientation, while valuable for well-defined challenges with clear solutions, is inadequate for compossibility crises, where the very framing of 'the problem' is contested and where solutions for one dimension create problems in others.

Crealectics shifts from problem-solving to what might be termed 'compossibility-creating', expanding the space of imaginable and enactable shared futures rather than optimising within predetermined constraints. This does not mean ignoring real difficulties or constraints, but recognising that constraints are partly constructed through how we frame situations. As Gregory Bateson (2000) observed, how we define problems significantly determines which solutions seem possible; changing our framing changes the landscape of compossibility.

This shift has important implications for how we understand obstacles to change. In problem-solving mode, resistance is typically framed as something to be overcome, with stakeholders portrayed as 'not understanding' the solution or prioritising their 'narrow interests' over the 'collective good'. In compossibility-creating mode, resistance becomes information about incompossibilities we may have failed to recognise, signalling that our proposed innovations are incompossible with values, identities, or needs we have excluded from consideration.

The Consciousness Dimension of Change

Throughout this chapter, I have emphasised that compossible innovation requires a shift in consciousness, not merely technical methods. Navigating towards compossible futures requires inner development, or innernovation, as much as outer action.

This emphasis on inner growth might appear to place responsibility for systemic change primarily at the individual level, potentially obscuring structural factors that constrain possibilities regardless of individual consciousness. This concern is worth taking seriously. Systems of power, accumulated privilege, embedded infrastructure, and path-dependent institutional arrangements profoundly shape what changes seem possible, often favouring the *status quo* and those who benefit from it (Scott, 1998).

Yet the relationship between consciousness and structure is creatively dialectical rather than unidirectional. Structures shape consciousness; our mental models, perceptual habits, and sense of possibility are profoundly influenced by the systems we inhabit. But consciousness also shapes structures: systems persist, in part, through the collective beliefs, assumptions, and practices that constitute them.

The crealectic emphasis on consciousness recognises that systemic change requires transforming not only policies, technologies, or institutions but also the paradigms from which these emerge. Donella Meadows (1999) identified 'paradigm change' as the highest-leverage point for system transformation, more powerful than changing rules, feedback loops, or even system goals, because paradigms determine how we interpret all these other elements. The crealectic method is part of a larger framework, C.I.P.H.E.R., which emphasises eudynamic balance between the Personal dimension and three other dimensions: Earth, Work, and Society (de Miranda, 2025).

Consciousness transformation in crealectic practice involves cultivating several capacities:

Presence: The ability to attend fully to present-moment reality rather than being lost in conceptual abstractions or mental narratives about situations. Mindfulness practices from contemplative traditions, adapted for organisational contexts, enhance this capacity (Kabat-Zinn, 1994).

Perspective-taking: The ability to genuinely understand how situations appear from diverse viewpoints, not merely intellectually but experientially. This differs from both projection (assuming others see as we do) and relativism (treating all views as equally valid regardless of evidence). It involves 'imaginative variation' (Husserl, 1970) and systematically explores how situations appear from different positions.

Systems sensing: The ability to perceive patterns, relationships, and feedback loops rather than just isolated events. This involves training attention to see wholes rather than just parts, processes rather than just states, relationships rather than just entities (Senge, 1990).

Future-oriented intentionality: The ability to sense and actualise emerging possibilities rather than merely reacting to current circumstances or defending past patterns. Scharmer's (2009) concept of 'presencing', operating from awareness of the highest future possibility rather than habitual patterns, captures this capacity.

Resilient vulnerability: The ability to remain open to uncertainty, ambiguity, and not-knowing without collapsing into premature closure or paralysing confusion. This involves being comfortable with uncertainty, which is essential for creativity.

These capacities are not personality traits individuals either possess or lack, but practices that can be cultivated through deliberate effort, both individually through contemplative practices and collectively through organisational learning processes.

Power, Privilege, and Compossibility

A critical question for any methodology of societal change concerns power asymmetries and how they shape which values, interests, and concerns are recognised as legitimate. If compossibility means ensuring diverse goods can coexist, whose goods count? Who decides which incompossibilities must be addressed and which can be dismissed?

These are not peripheral concerns but central to crealectic practice. The method could be co-opted to generate illusory consensus that serves powerful interests while marginalising vulnerable populations under the rhetoric of so-called win-win solutions. Indeed, much corporate sustainability discourse functions this way: it claims to reconcile economic growth with environmental protection through 'green growth' or 'sustainable development' that preserves fundamentally unsustainable systems with minor modifications (Swyngedouw, 2010).

Crealectic practice addresses this concern through several principles:

Explicit attention to power: Rather than pretending that dialogue occurs on level playing fields, crealectic processes may explicitly acknowledge power asymmetries and work to create conditions in which less powerful voices can genuinely influence outcomes. This might involve compensating community representatives for their time and expertise, providing technical support so communities can engage with complex issues on equal footing, establishing decision-making processes that prevent wealthy or powerful stakeholders from dominating through resource advantages, and creating accountability mechanisms to ensure commitments are honoured.

Centering affected communities: Compossibility assessments prioritise the experiences and concerns of those most directly affected by proposed changes rather than privileging distant experts or powerful stakeholders. Environmental justice principles, emphasising that 'the people closest to the problem are closest to the solution', may inform this approach (Bullard, 1990).

Historical awareness: Compossibility judgments must account for historical contexts that shape present inequities. What appears to be a fair compromise today may perpetuate historical injustices if it fails to address accumulated advantages and disadvantages. Repair and restoration, rather than merely avoiding future harm, become relevant to genuine compossibility (Coates, 2014).

Attention to non-human stakeholders: Compossibility must extend beyond human interests to include the wellbeing of ecosystems, species, and Earth systems that cannot represent themselves in human deliberations. This requires 'thinking like a mountain' (Leopold, 1949), expanding moral consideration beyond human concerns and immediate timeframes.

Structural transformation over superficial accommodation: Genuinely compossible innovation often requires transforming systems that currently generate incompossibility rather than merely accommodating diverse interests within unchanged systems. The energy transition probably entails transforming energy systems themselves.

Limitations and Appropriate Applications

While the crealectic method offers valuable frameworks for navigating compossibility challenges, it may not be universally applicable or sufficient for all forms of societal change. Recognising its limitations helps clarify appropriate applications.

Time and resource intensive: Crealectic processes require substantial time and resources for dialogue, exploration, prototyping, and iteration. In genuine emergencies requiring an immediate response, this deliberative approach may be inappropriate. Metaphorically speaking, when a building is burning, we need rapid evacuation, not extensive conversation about diverse evacuation preferences. However, apparent urgency can be manufactured to justify bypassing deliberation (Klein, 2007). Claims that 'there is no time for process' often serve powerful interests seeking to implement predetermined solutions without democratic scrutiny. Distinguishing genuine emergencies from manufactured urgency is itself an important discernment.

Requires baseline conditions: Crealectic practice assumes certain baseline conditions, such as stakeholders willing to engage in good faith, social infrastructure that enables dialogue, institutional contexts that allow experimentation, and resource sufficiency that enables exploration. Where these conditions are absent, for instance in contexts of active violence, extreme scarcity, or totalitarian control, other approaches may be necessary. The method works best for navigating complex trade-offs among stakeholders who share goodwill and basic commitments to finding compossible solutions, not for addressing situations where dishonest parties seek domination regardless of others' wellbeing.

Not a substitute for other change modes: Some situations require advocacy and resistance, rather than (or alongside) collaborative innovation. When systems are fundamentally unjust or destructive, working within them to find compossible accommodations may legitimise what should be opposed. Crealectic practice complements, but does not replace, social movements, political advocacy, legal challenges, and direct action that challenge rather than accommodate unjust systems.

Cultural specificity: The method articulated here arises from particular cultural contexts (Western philosophy, organisational development, design thinking) that may not translate directly to other contexts. While I believe the core insights about compossibility have broader validity, specific practices need adapting to local contexts, cultural norms, and indigenous knowledge systems. Imposing Western-derived methods as universal solutions would itself constitute a form of incompossibility.

Towards a Regenerocene

The ultimate horizon of crealectic practice is a 'Regenerocene': an era characterised not merely by reducing harm but by actively restoring and regenerating the ecological, social, and cultural systems that industrial civilisation has depleted. This vision, drawing from indigenous wisdom traditions, permaculture, and regenerative design, recognises that genuine flourishing requires healing past harms while creating conditions for future thriving (Mang & Reed, 2012).

The Regenerocene marks a shift from viewing humans as inherently destructive forces to be constrained through regulation and sacrifice, towards recognising human potential for participation in planetary healing and creativity. It acknowledges humanity's genuine capacity for innovation while channelling that capacity towards regeneration rather than endless extraction and accumulation.

Crealectics contributes to this transition by offering a methodological pathway for co-creating innovations through approaches that leave systems more eudynamic. What constitutes eudynamia and how to assess it remain research projects to be refined, but we can already say that it means creating energy systems that enhance rather than deplete communities, distribute rather than concentrate power, work with rather than against natural processes, and allow for what we might call personal flowrishing.

Yet, the Regenerocene cannot be merely technological or methodological. It requires a new dimension, a transformed understanding of who we are, how we relate to Earth and each other. This constitutes progress and flourishing. The crealectic emphasis on explorers of consciousness acknowledges that this transformation is ultimately a journey into new ways of becoming, not merely new ways of doing.

We are not detached observers on the shore, deciding where others should sail, but fellow voyagers learning together how to navigate waters none of us have crossed before. The challenges we face (climate disruption, biodiversity collapse, inequality, polarisation, fragmentation of meaning, a general compossibility crisis) have no precedent; their solutions will not be found in any historical playbook. They require us to become explorers willing to venture into unknown seas of consciousness and collaboration, discovering as we go what it means to enact change that honours the deep intercreation between all forms of life.

References

Abram, D. (1996). *The spell of the sensuous: Perception and language in a more-than-human world*. Pantheon Books.
Aldridge, D. (1996). *Music therapy research and practice in medicine: From out of the silence*. Jessica Kingsley Publishers.
Alexander, C. (2002). *The nature of order: An essay on the art of building and the nature of the universe* (Books 1–4). Center for Environmental Structure.
Bachelard, G. (1958). *La poétique de l'espace*. Presses Universitaires de France. [English translation: *The poetics of space*, M. Jolas, Trans., 1964, Orion Press].
Barabási, A.-L. (2002). *Linked: The new science of networks*. Perseus Publishing.
Bateson, G. (2000). *Steps to an ecology of mind*. University of Chicago Press.
Benyus, J. M. (1997). *Biomimicry: Innovation inspired by nature* (Rev. ed.).
Bergson, H. (2007). *Creative evolution* (A. Mitchell, Trans.). Palgrave Macmillan. (Original work published 1911).
Bohm, D. (1996). *On dialogue*. Routledge.
Breton, A. (1969). Manifestoes of surrealism (R. Seaver & H. R. Lane, Trans.). University of Michigan Press.
Brown, T. (2009). *Change by design: How design thinking transforms organizations and inspires innovation*. HarperBusiness.
Bruner, J. (1990). *Acts of meaning*. Harvard University Press.
Bullard, R. D. (1990). *Dumping in Dixie: Race, class, and environmental quality*. Westview Press.
Cajete, G. (2000). *Native science: Natural laws of interdependence*. Clear Light Publishers.
Capra, F., & Luisi, P. L. (2014). *The systems view of life: A unifying vision*. Cambridge University Press.
Carson, S. H. (2011). Creativity and psychopathology: A shared vulnerability model. *Canadian Journal of Psychiatry, 56*(3), 144–153.
Chambers, R. (1997). *Whose reality counts?* Intermediate Technology Publications.
Coates, T.-N. (2014). *The case for reparations*. The Atlantic.
de Bono, E. (1970). *Lateral thinking: Creativity step by step*. Harper & Row.
de Miranda, L. (2025). *Crealectics as creative method: A new approach to holistic innovation for compossible worlds*. Palgrave Macmillan.

Descartes, R. (1999). *Discourse on method* (D. A. Cress, Trans.). Hackett Publishing. (Original work published 1637).
Escobar, A. (2018). *Designs for the pluriverse: Radical interdependence, autonomy, and the making of worlds*. Duke University Press.
Gendlin, E. T. (1978). *Focusing*. Everest House.
Guilford, J. P. (1967). *The nature of human intelligence*. McGraw-Hill.
Heidegger, M. (1977). *The question concerning technology and other essays* (W. Lovitt, Trans.). Harper & Row.
Holland, J. H. (2014). *Complexity: A very short introduction*. Oxford University Press.
Holling, C. S. (1978). *Adaptive environmental assessment and management*. John Wiley & Sons.
Husserl, E. (1970). *The crisis of European sciences and transcendental phenomenology* (D. Carr, Trans.). Northwestern University Press.
Jung, C. G. (1968). The archetypes and the collective unconscious (2nd ed., R. F. C. Hull, Trans.). Princeton University Press.
Jung-Beeman, M., Bowden, E. M., Haberman, J., Frymiare, J. L., Arambel-Liu, S., Greenblatt, R., Reber, P. J., & Kounios, J. (2004). Neural activity when people solve verbal problems with insight. *PLoS Biology, 2*(4), e97.
Kabat-Zinn, J. (1994). *Wherever you go, there you are: Mindfulness meditation in everyday life*. Hyperion.
Kauffman, S. (1995). *At home in the universe: The search for the laws of self-organization and complexity*. Oxford University Press.
Kauffman, S. (2019). *A world beyond physics: The emergence and evolution of life*. Oxford University Press.
Kimmerer, R. W. (2013). *Braiding sweetgrass: Indigenous wisdom, scientific knowledge, and the teachings of plants*. Milkweed Editions.
Klein, J. T. (1990). *Interdisciplinarity: History, theory, and practice*. Wayne State University Press.
Klein, N. (2007). *The shock doctrine: The rise of disaster capitalism*. Metropolitan Books.
Lakoff, G., & Johnson, M. (1999). *Philosophy in the flesh: The embodied mind and its challenge to Western thought*. Basic Books.
Leibniz, G. W. (1989). *Philosophical essays* (R. Ariew & D. Garber, Trans.). Hackett publishing. (Original work published 1714).
Leopold, A. (1949). *A sand county almanac*. Oxford University Press.
Mang, P., & Reed, B. (2012). Designing from place: A regenerative framework and methodology. *Building Research & Information, 40*(1), 23–38.
March, J. G. (1991). Exploration and exploitation in organizational learning. *Organization Science, 2*(1), 71–87.
McClintock, B. (1984). The significance of responses of the genome to challenge. *Science, 226*(4676), 792–801.

Meadows, D. (1999). *Leverage points: Places to intervene in a system*. The Sustainability Institute.
Meadows, D. (2008). *Thinking in systems: A primer*. Chelsea Green Publishing.
Mouffe, C. (2000). *The democratic paradox*. Verso.
Mueller, P. A., & Oppenheimer, D. M. (2014). The pen is mightier than the keyboard: Advantages of longhand over laptop note taking. *Psychological Science, 25*(6), 1159–1168.
Newen, A., De Bruin, L., & Gallagher, S. (Eds.). (2018). *The Oxford handbook of 4E cognition*. Oxford University Press.
Odum, E. P. (1969). The strategy of ecosystem development. *Science, 164*(3877), 262–270.
Osborn, A. F. (1953). *Applied imagination: Principles and procedures of creative problem-solving*. Charles Scribner's Sons.
Paulus, P. B., & Nijstad, B. A. (Eds.). (2003). *Group creativity: Innovation through collaboration*. Oxford University Press.
Perrow, C. (1999). *Normal accidents: Living with high-risk technologies*. Princeton University Press.
Piketty, T. (2014). *Capital in the twenty-first century* (A. Goldhammer, Trans.). Harvard University Press.
Pressman, J. L., & Wildavsky, A. (1984). *Implementation* (3rd ed.). University of California Press.
Rockström, J., Steffen, W., Noone, K., Persson, Å., Chapin, F. S., Lambin, E. F., Lenton, T. M., Scheffer, M., Folke, C., Schellnhuber, H. J., Nykvist, B., de Wit, C. A., Hughes, T., van der Leeuw, S., Rodhe, H., Sörlin, S., Snyder, P. K., Costanza, R., Svedin, U., et al. (2009). A safe operating space for humanity. *Nature, 461*(7263), 472–475.
Root-Bernstein, R., & Root-Bernstein, M. (1999). *Sparks of genius: The thirteen thinking tools of the world's most creative people*. Houghton Mifflin.
Sawyer, R. K. (2012). *Explaining creativity: The science of human innovation* (2nd ed.). Oxford University Press.
Scharmer, C. O. (2009). *Theory U: Leading from the future as it emerges*. Berrett-Koehler Publishers.
Schein, E. H. (2010). *Organizational culture and leadership* (4th ed.). Jossey-Bass.
Schön, D. A. (1983). *The reflective practitioner: How professionals think in action*. Basic Books.
Schwartz, P. (1991). *The art of the long view: Planning for the future in an uncertain world*. Doubleday.
Scott, J. C. (1998). *Seeing like a state: How certain schemes to improve the human condition have failed*. Yale University Press.
Senge, P. M. (1990). *The fifth discipline: The art and practice of the learning organization*. Doubleday.

Snowden, D. J., & Boone, M. E. (2007). A leader's framework for decision making. *Harvard Business Review, 85*(11), 68–76.

Steffen, W., Richardson, K., Rockström, J., Cornell, S. E., Fetzer, I., Bennett, E. M., Biggs, R., Carpenter, S. R., de Vries, W., de Wit, C. A., Folke, C., Gerten, D., Heinke, J., Mace, G. M., Persson, L. M., Ramanathan, V., Reyers, B., & Sörlin, S. (2015). Planetary boundaries: Guiding human development on a changing planet. *Science, 347*(6223), 1259855.

Suzuki, S. (1970). *Zen mind, beginner's mind: Informal talks on Zen meditation and practice*. Weatherhill.

Swyngedouw, E. (2010). Apocalypse forever? Post-political populism and the spectre of climate change. *Theory, Culture & Society, 27*(2–3), 213–232.

Tufekci, Z. (2017). *Twitter and tear gas: The power and fragility of networked protest*. Yale University Press.

Turkle, S. (2011). *Alone together: Why we expect more from technology and less from each other*. Basic Books.

Varela, F. J., Thompson, E., & Rosch, E. (1991). *The embodied mind: Cognitive science and human experience*. MIT Press.

Verger, N. B. (2026) (Chapter 2, this volume). Social-(ecological) innovation as a scaled-form of creative preservation. *Possibility Studies & Societal Futures: Imagination, Change, and Preservation*.

Verger, N. B., Duymedjian, R., & Glăveanu, V. P. (2025). The meritocracy of preservation: Reimagining merit beyond production. *Organization, 33*. https://doi.org/10.1177/13505084251347519

Verger, N. B., Duymedjian, R., Wegener, C., & Glăveanu, V. P. (2024). Creative preservation: A framework of creativity in support of degrowth. *Review of General Psychology, 28*(3), 268–281.

Walters, C. J. (1986). *Adaptive management of renewable resources*. Macmillan.

Watts, A. W. (1975). *Tao: The watercourse way*. Pantheon Books.

Wenger, E. (1998). *Communities of practice: Learning, meaning, and identity*. Cambridge University Press.

Whitehead, A. N. (1978). Process and reality (Corrected ed.). Free Press. (Original work published 1929).

CHAPTER 4

Can an Old Dog Teach New Tricks? Older People as Agents of "Slow Futurition"

Francesca Lorenzi

Introduction

The contemporary era is often described as one marked by the convergence of global challenges—environmental crises, socio-political instability, and rapid technological disruption. Alvin Toffler (2022), in Future Shock, frames this predicament as the result of an "accelerative thrust" that propels society through an increasingly compressed experiential channel (ibid., p. 35). Toffler argues that this accelerated pace of change fosters a pervasive sense of "transience"—a feeling of impermanence that permeates every aspect of life. While these rapid changes are often framed as inevitable, they also reveal a paradox: the forces driving innovation and societal growth may contribute to the emergence of an unsustainable economic model, sustained by perpetual forward momentum and the relentless pursuit of novelty.

At the heart of this paradox lies the imperative for constant innovation. The pressure to innovate compels societies to exploit existing resources

F. Lorenzi (✉)
Centre for Possibility Studies, Dublin City University, Dublin, Ireland
e-mail: francesca.lorenzi@dcu.ie

N. B. Verger, D. Mattson (eds.), *Possibility Studies and Societal Futures*, Palgrave Advances in Possibility Studies,
https://doi.org/10.1007/978-3-032-26379-7_4

while continually seeking new solutions. However, this innovation-driven model is ultimately unsustainable. Toffler's analysis underscores the tension between society's need for progress and the risks inherent in such an economic system. In this context, the pursuit of novelty can deplete resources, exacerbating existing challenges in sustainability and equity.

Verger et al. (2024) argue that the global challenges we face also present an opportunity to reimagine both our approach to sustainability and the social roles of marginalised groups—particularly those excluded from dominant, growth-oriented discourses of productivity and disruptive innovation. In response, they introduce the concept of "creative preservation" as an alternative to the exploitative, growth-centric paradigm. This framework emphasises care for what already exists, advocating a cultural and ethical shift that values sustainability and the conservation of both material resources and knowledge. Rather than promoting perpetual consumption and exploitation, creative preservation fosters long-term, meaningful engagement with resources, ideas, and practices, encouraging reflection on continuity, care, and interdependence in navigating an uncertain world. Creative preservation "grounds human intention and action within the context of an ongoing and mutually constitutive engagement between people and the environment" (Verger et al., 2024, p. 269), where continuity across temporal dimensions helps sustain what is worthwhile in a world dominated by change.

Building on this foundation, the chapter introduces a broader concern with temporality and the contributions of older generations to future-making. Recognising the interdependence of past, present, and future makes it possible to revalue older individuals not as relics of a bygone era but as active custodians of memory and meaning-making. This perspective challenges dominant narratives that portray old age as a stage of decline—focused on retrospection rather than potential for innovation. Instead, it posits that the interplay between past and future can yield critical insights for navigating contemporary challenges, positioning older adults as sources of resilience, continuity, and creativity.

Cultural representations of old age often oscillate between two limiting stereotypes: the infantilised elder and the venerable sage. Both obscure the dynamic potential of later life. To avoid reproducing ageist or exclusionary assumptions in possibility studies and future thinking, it is essential to interrogate these narratives and reveal their underlying temporal biases. As Zittoun (2022) notes, "old age may be a period in which the possible plays a unique role" (ibid., p. 5). This insight invites a re-conceptualisation of older adulthood as a site of ongoing engagement, transformation, and creative potential. Accordingly, this chapter advocates a temporally complex understanding of future-making that recognises older adults as active agents in processes of innovation, critical inquiry, and intergenerational exchange.

To advance this objective, the discussion proceeds along two interrelated lines of analysis: first, it examines how restrictive, linear conceptions of time reinforce ageist tropes equating youth with innovation and old age with obsolescence; second, it critiques developmental models—such as Erikson's (1950, 1959, 1982) theory of generativity—that, while emphasising future orientation, may inadvertently devalue the accumulated knowledge of older adults. Drawing on Glăveanu's (2022) relational understanding of the possible and the notion of future heritage, the chapter argues for futures that are both relationally and temporally distributed. Here, the past is not a static repository of outdated knowledge but a vital resource for shaping more inclusive and participatory futures (Sandford, 2019).

Although contextualisation may risk fostering parochialism, it provides a necessary foundation for cultivating a more harmonious future—one that progressively broadens the circle of moral concern beyond the immediate needs of individual communities (Singer, 1981). Past experiences that once prompted exploration of particular possibilities often remain embedded in memory, shaping the practices we enact in the present. These practices, in turn, may open pathways for investigating new or seemingly unrelated possibilities, thereby generating further opportunities for future inquiry. Simultaneously, the perspectives cultivated through such exploration can prompt a re-examination of the past through newly acquired interpretive lenses, offering renewed insights into experiences once regarded as familiar (Verger et al., 2024). In this way, the possible functions as a connective thread that weaves together different dimensions of temporality, moving fluidly and multi-directionally across past, present, and future.

Finally, the chapter introduces the interconnected concepts of socio-cultural generativity and slowed-down futurition—a deliberate, reflective mode of future thinking that resists the accelerating tempo of contemporary life. Slowed-down futurition (Eshel, 2012) encourages inclusive, multi-temporal engagement with the possible, reframing futurity not as relentless progress but as a space for deliberation, reciprocity, and care. Consistent with the ethos of creative preservation (Verger et al., 2024), this approach underscores that certain groups—particularly older adults who embody the permanence of the past within the present—can contribute to reimagining sustainable ways of flourishing by exploring and recombining the possibilities inherent in what already exists.

Rethinking Temporal Frameworks

To fully appreciate the contributions of older generations to possible futures, it is essential to interrogate stereotypes rooted in narrow temporal frameworks. A necessary starting point for this interrogation is an analysis of time itself and of how its theorisation shapes conceptions of human development.

Temporal framing plays a pivotal role in shaping perceptions of reality, particularly by legitimising and naturalising social norms and expectations. As Lazar (2019) argues, temporal framing is a rhetorical strategy frequently employed in constitutional preambles to "sell the future" by invoking and reconfiguring the authority of the past. This legitimation is anchored in storytelling practices and narrative structures that actively construct causal relationships between the past, present, and imagined futures (ibid., p. 3). Lazar (2019) further underscores that "meaning depends on narrative structure, and narrative structure depends on a conception of the flow of time", which may unfold according to linear, cyclical, or episodic patterns (ibid., p. 5). In this way, temporal framing establishes legitimacy by delineating acceptable forms of order, sanctioned modes of action, and recognisable patterns of relationality within specific temporal configurations.

Dominant linear conceptualisations of time tend to oscillate between two poles. On one end lies primitivism, which critiques modern civilisation by idealising the past and envisioning the future as a revival of its perceived virtues—such as simplicity, authenticity, and harmony with nature (Malendowicz, 2019). On the other end stands modernist progressivism, which enshrines the future as the primary locus of development, innovation, and advancement. Situated between these extremes is the degrowth or post-growth conceptualisation of temporality, which dissociates the future from the logic of perpetual economic expansion (Verger et al., 2024; 2025). Instead, it grounds futurity in the creative and sustainable re-engagement with existing resources, thereby emphasising the dynamic interplay of all temporal dimensions—past, present, and future—within an integrated and regenerative temporal framework (see also De Miranda, Chap. 3, this volume).

A strictly linear conception of time, understood as an objective, chronological continuum, also imposes significant limitations. It risks constraining understandings of progression and obscuring alternative temporal modalities—such as fragmented, cyclical, or subjective time—that may offer more nuanced insights. Furthermore, time is not universally experienced as linear; rather, it is interpreted and measured differently across cultural contexts. Facer (2023) underscores the dominance of certain "timescapes"—such as Anthropocene or gendered timescapes—that eclipse lived experiences of time, thereby reinforcing hegemonic perspectives while marginalising alternative temporalities.

Subjective time—understood as "the conscious or unconscious mental representation of physical temporal information" (Thönes & Stocker, 2019, p. 114)—resists the linear logic of temporal succession (Strawson, 2004) and is instead perceived as fluid and contextually contingent (Nadobnik et al., 2021). In this view, time is best understood as temporality—a mode in which images of the past and anticipations of the future coexist in the present, shaped by the evolving circumstances of experience. Given the inherently retrospective nature of sense-making (Weick, 1979; Weick et al., 2005), individuals constantly reinterpret the past in light of present events, thereby constructing evolving orientations towards the past, present, and future (Zimbardo & Boyd, 2008). Ageing, in this regard, constitutes a specific temporal frame shaped by lived experience, wherein diverse "temporal horizons blend" (Massie & Staude, 2024, p. 573). Past experiences mediate older adults' engagement with the present, functioning as interpretive frameworks through which they imagine and negotiate future possibilities. For example, an individual with a lifelong history of community activism may approach present opportunities for volunteering not merely as leisure but as a continuation of a meaningful trajectory, thereby envisioning the future in terms of ongoing contribution.

Henri Bergson's (1910) concept of durée (duration) offers a further basis for the critique of deterministic models of time. He contends that time, in its essence, is not a sequence of uniform units but a qualitative, non-linear flow apprehended directly by consciousness. In contrast to spatialised time—which fragments and measures experience—durée unfolds like a melody, with each moment gaining meaning in relation to the whole (Bergson, 1910). Past experiences are not left behind but interpenetrate the present, continually shaping perception and thought (Guerlac, 2006). This perspective disrupts mechanistic understandings of time and foregrounds the centrality of memory and subjective awareness in the construction of experience (Mullarkey, 2007).

Sense-making is inherently backward-looking (Weick, 1979; Weick et al., 2005); therefore, a meaningful envisioning of the future necessarily engages with the past. Facer (2024) warns that temporal frames that dissociate future narratives from those of the past "often produce dissonance and conflict" (ibid., p. 3). She argues that timing practices act as "a powerful force in structuring the perception of possibility" (ibid., p. 3). Presuming a singular, objective measure of time risks marginalising diverse temporal perspectives and thereby constraining our collective capacity to imagine plural and inclusive futures. To mitigate this risk, Facer (2024) calls for cultivating a "temporal imagination" that disrupts dominant assumptions and opens "new pathways to alternative worlds", enabling narratives that reflect multiple ways of knowing and being (ibid., p. 3).

Crucially, the selection of timing practices reflects and reproduces dominant cultural and political values, producing tangible material, social, and symbolic effects (Facer, 2024, p. 2). For example, structuring the school day into rigid timetables privileges efficiency, productivity, and standardisation—values aligned with industrial and bureaucratic logics—while marginalising alternative temporalities, such as pauses for reflection, communal rhythms, or intergenerational exchange.

The strategic deployment of temporal frames can exacerbate societal divides and entrench existing inequalities (Facer, 2024, p. 3). Nadobnik et al. (2021) propose a metaphor to conceptualise these dynamics: temporality as a ribbon composed of interwoven threads that together form the broader social fabric. In this view, altering the configuration of the cloth does not affect individual fibres in isolation; rather, it reconfigures the broader pattern of collective life—both intentionally and unintentionally (Nadobnik et al., 2021, p. 10).

Recognising these multiple temporal frameworks requires rethinking linear, deterministic views of time. It underscores the importance of subjective, lived, and situated temporalities in shaping human experience. Furthermore, establishing non-linear connections among all temporal dimensions ensures "continuity of objects and ideas" (Verger et al. 2024, p. 269). In particular, within a meritocracy of preservation (Verger et al., 2025), value is not derived from creating the new but from the capacity to sustain, repair, and creatively re-engage with what already exists.

Preservation thus becomes an imaginative act—a form of care that transforms continuity into a site of possibility rather than limitation. This reorientation of value systems invites a more inclusive understanding of human agency across the life course, acknowledging that the ability to nurture, reinterpret, and maintain is as vital as the ability to innovate.

Heritage, in this light, emerges not as a constraint but as a generative resource for constructing alternative futures grounded in experience, continuity, and relationality.

Beyond Stereotyping of Older Age

Like other forms of prejudice and discriminatory behaviour, ageism is deeply rooted in personal insecurity and a fear of what is perceived as unfamiliar or "other". Butler (1969, p. 243) characterises ageism as a form of bigotry that "reflects a deep-seated uneasiness on the part of the young and middle-aged—a personal revulsion to and distaste for growing old, disease, disability; and fear of powerlessness, 'uselessness', and death". What sets ageism apart from other prejudices is its fundamental irony: the very process that incites such discriminatory attitudes—ageing—is a universal and inescapable human experience that, provided longevity, will affect those who perpetuate these biases. In this sense, ageism can be understood as a denial of the temporal nature of human existence and an evasion of the existential realities associated with the passage of time.

Engaging with ageism requires treating temporality not merely as a chronological framework but as a multidimensional structure that shapes how ageing is experienced and represented. Massie and Staude (2024) argue that "the temporality of ageing presents an irreducible complexity", comprising biological time, personal narrative time, and historical time (p. 565). Recognising these interwoven temporalities offers a more comprehensive framework for understanding ageing, one that disrupts reductive developmental paradigms and challenges dominant cultural narratives. As Luppi (2015) explains, the lifespan perspective introduced by Baltes et al. (1980) reconceptualised human development as dynamic, non-linear, and discontinuous. This model fundamentally challenges the conventional trajectory of linear ascent followed by inevitable decline, instead foregrounding the variability, complexity, and heterogeneity that characterise individual life courses (Luppi, 2015, p. 158).

Nevertheless, despite the theoretical innovation and conceptual breadth this perspective offers, it has not fully displaced the long-standing developmental preoccupation with childhood and youth, which continues to dominate mainstream narratives of human growth, maturity, and change. In response to this enduring imbalance, Butler (1969) famously referred to older adulthood as "the neglected stepchild of human development", thereby illuminating the persistent marginalisation of later life within developmental and psychological discourse.

This marginalisation is further perpetuated through what Gullette (2018) identifies as the "hegemonic decline ideology", a cultural framework that constructs ageing primarily as an unavoidable process of deterioration and loss. Such deficit-oriented representations cast older adults as fundamentally "other", thereby legitimising their exclusion from full participation in social, political, and cultural life. Manifestations of this exclusion are evident across multiple domains, including workplace ageism, the digital divide—which for many older adults translates into restricted access to essential services—and broader systemic invisibility, all of which collectively reinforce age-based inequities.

Thornton (2002) identifies six dominant stereotypes of ageing—poor health and disability, cognitive deterioration, emotional distress, lack of sexuality, inevitable decline in vitality, and an inability to learn or be productive. Among these stereotypes, the notion that older adults lack future--oriented thinking is particularly damaging, as it sustains cultural narratives that undermine their agency and social relevance. This perception, however, arises less from any genuine absence of future orientation than from a dominant, linear conceptualisation of time that narrowly defines what counts as "the future". When viewed through the lens of past experiences and accumulated life histories—perspectives more commonly associated with later life—older people's temporal orientations can be understood as a distinctive form of future thinking. Although it departs from dominant linear narratives, this mode of temporal imagination nevertheless constitutes a meaningful and generative engagement with the future.

These reductive conceptualisations are not new. Classical texts have long cast ageing as decline. In *De Senectute*, Cicero (1923) associates old age with frailty, withdrawal, and proximity to death, while Aristotle pathologises it as bodily decay (On the Soul 1936,108b 18–29; 1985, Problems 861a 25–29). These early portrayals framed ageing as a departure from the idealised past and helped shape a cultural logic in which the present is a space of loss and degeneration. As Gullette (2018, p. 254) notes, these enduring portrayals reinforce Butler's assertion that ageism is grounded in cultural discomfort with loss, impairment, and mortality.

More recently, media portrayals of the "young old"—active, vibrant, and self-sufficient—have sought to challenge ageist stereotypes. However, these representations often reinforce exclusion by setting aspirational, youth-centric standards that marginalise those who do not conform. As Jonson (2012, p. 198) observes, such idealised images risk legitimising age-based discrimination by disguising it as celebration. In doing so, they perpetuate the false binary between "successful" and "unsuccessful" ageing, obscuring the full range of experiences in later life.

Crucially, ageism intersects with ableist ideologies, further entrenching marginalisation. The biomedical framing of ageing links physical and cognitive impairments to diminished human worth. Thomas (1999, p. 133) reminds us that "what is and what counts as impairment is always socially located", revealing that both ageism and ableism rest on normative expectations of autonomy and productivity. The conflation of ageing with inevitable decline erodes recognition of older people's capabilities and leads to the widespread assumption that impairment equates to a lack of agency.

This is especially evident in the infantilisation of older adults, which reflects broader societal discomfort with ageing. Ironically, older people are often treated in ways that mirror how children are treated. While both groups are viewed as dependent, children are imagined as adults-in-the-making, imbued with future potential (Holtorf & Bolin, 2024). By contrast, the dependency of older adults is framed as permanent. This distinction underpins practices such as "elderspeak" and patronising care routines in institutional settings, including nursing homes. Kemper and Harden (1999) argue that these behaviours—often unconscious—undermine dignity and contribute to both cognitive decline and psychological harm. Such infantilisation reveals how deeply deficit-based constructions of ageing are embedded in everyday interactions.

Yet these cultural scripts are increasingly challenged by research on cognitive ageing, which reveals a more nuanced picture. While some cognitive functions, particularly fluid intelligence—quick information processing and adaptability—do decline with age (Murman, 2015; Zaninotto et al., 2018), the extent of this decline is highly variable and influenced by medical, lifestyle, socioeconomic, and gender-related factors (Zaninotto et al., 2018). Other cognitive functions, especially crystallised intelligence—such as vocabulary, general knowledge, and numerical skills—tend to remain stable or even improve over time (De Brito et al., 2023). Khammash (2023) contends that crystallised intelligence encompasses cognitive abilities that rely heavily on accumulated knowledge and experience. It develops progressively through the continuous acquisition of information, skills, and cultural understanding across the lifespan. Consequently, the extensive reservoir of knowledge and expertise built over time often affords older adults a cognitive advantage in tasks that draw upon prior learning. In contrast, fluid intelligence involves cognitive flexibility, problem-solving ability, and the capacity to adapt knowledge to novel situations. Although fluid intelligence typically declines with age, this reduction is frequently offset by older individuals' ability to draw upon their rich repertoire of experiences. This experiential grounding enables them to interpret and respond to new circumstances with insight informed by past encounters.

Older adults also demonstrate retained procedural memory and increased capacity to filter distractions, often outperforming middle-aged adults in tasks requiring focused attention (Veríssimo et al., 2022). The concept of cognitive adaptability undermines the stereotype that older people cannot learn or think ahead. Even in the context of slower processing, older individuals can integrate new knowledge through reflective and deliberate strategies that build upon prior experience. As such, cognition-based ageism—like all stereotyping—obscures individual variability and tends to overemphasise loss while overlooking retained or adaptive strengths. According to Levy's (2009) Stereotype Embodiment Theory, internalised negative stereotypes can have tangible effects, including physical and mental health decline.

The assumption that cognitive impairment precludes creativity has contributed to the systematic devaluation of individuals living with dementia or other neurocognitive disorders. Camic et al. (2018, p. 9) challenge this deficit-based view by highlighting the often-overlooked creative capacities that persist within this population. This critique underscores a broader issue: the tendency to conflate ageing with cognitive decline, thereby obscuring alternative forms of contribution and expression. Models of "mini-c" and "little-c" creativity (Craft, 2000, 2001a, 2001b, 2003; Kaufman & Beghetto, 2009) are particularly relevant here, as they conceptualise creativity as grounded in everyday experiences rather than limited to exceptional or professional achievement. As Camic et al. (2018, p. 2) note, mini-c creativity involves internal processes of generating ideas and making connections that may remain invisible to others and are therefore frequently overlooked for their creative value.

Alongside deficit-based narratives, another reductive trope persists: the romanticisation of older people as inherently wise. Although this may seem positive, it imposes idealised expectations that obscure the realities of later life. Gordon and Jordan (2017) describe the tension between narratives of inevitable decline and those of transcendent wisdom, noting that the latter can be just as limiting. Even when older adults are viewed positively, it is often as the kindly sage—a stereotype that, while benevolent, still reduces complexity to an archetype.

Edmonson (2009) offers a more sophisticated view, defining wisdom not as an age-bound trait but as a relational, humanistic capacity rooted in acceptance and humility. In this conception, wisdom entails reflective and empathetic engagement with life's uncertainties. Machado de Oliveira (2021, p. xxiv) similarly rejects the myth of the perfected moral elder, writing that "grandmothers' gifts were crafted and offered through their mistakes, their paradoxes, and the complexities of their lives". It is through this embrace of imperfection that older individuals may offer valuable insight into the human condition. Wisdom, therefore, should be understood as a lived response to experience rather than a fixed attribute of ageing.

Beyond individual traits, broader cultural and economic systems also shape perceptions of ageing. Moscovici (1988) conceptualises ageing as a shared social representation, noting that attitudes towards ageing vary across contexts. Löckenhoff et al. (2009; 2022) suggest that more socioeconomically advanced societies tend to hold more negative views of ageing, likely due to neoliberal ideals that equate human worth with productivity. Within this framework, older adults are often seen as economically inactive and dependent, thereby reducing their perceived value (for a discussion on values and creative preservation, see Verger, Chap. 2, this volume).

Historically, the introduction of pension systems in Western countries in the late nineteenth century contributed to this perception. Costa (1998) argues that the concept of retirement became synonymous with economic withdrawal, reinforcing views of older people as passive and care-dependent (Datta Roy & Panda, 2022). Jolanki (2009) adds that retirees are often cast as recipients of action rather than autonomous agents. This marginalisation is further exacerbated by technological developments that render older adults' accumulated knowledge seemingly obsolete (Ayalon & Tesch-Römer, 2018).

These cumulative dynamics contribute to a gradual erosion of agency, undermining older people's status as capable social actors. However, the concept of "active ageing" offers a counter-narrative. Walker and Maltby

(2012) argue that the shift from wage-based employment to pension-based retirement often excludes people from broader civic life. They call for a redefinition of activity that encompasses not only economic participation but also social and political engagement (ibid., p. 118).

Active ageing, as promoted by WHO (2022), centres on maintaining functional ability—the capacity to do what one values. This includes mobility, decision-making, maintaining relationships, and contributing to society. Beard et al. (2016) and Rudnicka et al. (2020) emphasise that functional ability is shaped by both intrinsic capacity and environmental supports. Crucially, it must be distinguished from economic productivity alone. Preserving and promoting functional ability is essential not only for enhancing quality of life but also for safeguarding older adults' agency.

In sum, the cultural construction of ageing oscillates between deficit and idealisation—between assumptions of decline and projections of virtue. Both extremes restrict how ageing is socially imagined and institutionally treated. Confronting ageism, therefore, requires more than rejecting stereotypes. It demands a reconfiguration of the frameworks—biological, cultural, and temporal—through which ageing is understood.

Creativity in Older Adulthood

While biological ageing may constrain some aspects of procreative generativity, these limitations do not apply to creativity, which remains accessible and meaningful throughout the later stages of life. However, when generativity is narrowly framed within neoliberal paradigms that equate value with economic productivity, older adults are often portrayed as dependent, unproductive, or even regressive (Datta Roy & Panda, 2022). In contrast, a broader, ethically grounded conceptualisation of generativity reframes ageing as a stage rich in moral capacity and creative potential, rather than decline (Fried et al. (1997).

From this perspective, generativity is not contingent solely on physical vitality or cognitive sharpness but is rooted in a relational orientation towards care and contribution. This reconceptualisation is particularly salient in discussions of creativity in later life, which have traditionally been marginalised by cognitively reductionist frameworks that exclude individuals experiencing neurocognitive decline (Camic et al., 2018). Creativity in this context is often instrumentalised as therapeutic for older adults, rather than acknowledged as an intrinsic expression of theirs.

Although certain cognitive faculties, such as divergent thinking, may decline (Ross et al., 2023; McCrae et al. 1987) creative potential in older adulthood remains robust. Studies show that individuals with dementia can still produce meaningful artistic expression (Palmiero et al., 2012) and retain problem-solving abilities, even when cognitive flexibility is diminished (Ross et al., 2023). In fact, Kontos et al. (2013) have demonstrated that individuals with advanced dementia can make recognisable creative contributions, challenging dominant assumptions about cognitive deterioration. In a similar vein, Tromp and Glăveanu (2023) argue that creativity in later life involves not simply overcoming limitations but reframing them as conditions for new possibilities. Attitudes towards creativity, previous experiences, and one's interpretive stance all influence how older individuals engage with and leverage constraints in productive ways. Furthermore, when the contributions of older people are understood through the meritocratic lens of creative preservation—which emphasises "the generative potential of existing relations and the dignity of all things-in-the-world" (Verger et al., 2025, p. 8) and the necessity of building a robust axiological system underpinned by reciprocal respect and continuity across temporal dimensions and between humans and non-humans—older individuals emerge as essential conduits for sustaining, reinterpreting, and reimagining existing bodies of knowledge and practice.

Cognitively healthy older adults bring to creative pursuits a rich reservoir of life experience, accumulated knowledge, and linguistic sophistication (Whiting & Smith, 1997). While reliance on past experience may constrain radical innovation in some cases, it often fosters narrative openness—the capacity to reinterpret one's life in flexible and expansive ways (Gubrium, 1995; Zittoun, 2013). Zittoun (2023) highlights how imagination in later life often draws on a deep well of memory and meaning, enabling heightened engagement with what is possible. According to Hickson and Housley (1997), this inward turn towards creativity may be enhanced in older age, serving as a source of renewal and self-expression.

Creativity in this context fosters psychic wholeness, as Rugh (1991) notes, providing older individuals with a means of integrating lived experience into a coherent and transformative narrative. This process not only enriches personal meaning but also deepens social participation. Despite a reduction in available options, individuals retain potential projects and remain prospectively oriented (Massie & Staude, 2024), offering perspectives shaped by historical consciousness and sociocultural complexity.

Generativity in Later Life

Historically, psychological models of human development have privileged childhood and early adulthood as the principal arenas of growth and potential, often conceptualising maturity as the apex of physical, cognitive, and productive capacity. Within this framework, later life has frequently been cast as a period of decline—biologically, cognitively, and socially. However, contemporary scholarship increasingly contests this linear and hierarchical model, emphasising that older adulthood can be a period of ongoing development, adaptability, and contribution. This shift invites a more nuanced appreciation of ageing as a phase that extends, rather than concludes, the human potential for creativity, generativity, and meaning-making.

In his seminal work *Childhood and Society*, Erikson (1950) proposed an eight-stage psychosocial model of development grounded in the epigenetic principle, which holds that human growth unfolds in a sequence of stages, each building on the previous one. Central to this model is the concept of generativity, situated in the seventh stage of development—typically associated with middle adulthood (approximately ages 40–65). Erikson (1950) defined generativity as a concern for establishing and guiding the next generation, encompassing procreativity, productivity, and creativity (p. 267). By contrast, failure to achieve generativity results in stagnation, a psychological condition characterised by self-absorption and disconnection from broader social responsibilities.

While Erikson acknowledged that psychosocial development continues across the lifespan, his original formulation placed generativity firmly within middle adulthood, implicitly limiting its relevance for older adults. This limitation was addressed in the revised edition of *Childhood and Society*, co-authored with Joan Erikson in 1997, which added a ninth stage of development to account for the psychological tasks encountered in late adulthood (typically ages 80 and beyond). In this extension, the Eriksons introduced the concept of grand-generativity—a revitalised form of generativity that involves continued engagement in caregiving, mentoring, and the transmission of values to future generations.

Importantly, while the inclusion of this ninth stage acknowledges the enduring developmental potential of later life, the prefix grand may inadvertently reinforce the marginalisation of older adults by implying a derivative or secondary form of generativity. A more integrative approach would reconceptualise generativity as a lifelong, dynamic process that ebbs and flows across different phases of life. From this perspective, generativity should not be bound to chronological age but understood as a capacity for relational, moral, and cultural investment that persists throughout the life course (Manheimer, 1995, p. 14).

Building on Erikson's foundational insights, Kotre (1984) proposed a fourfold typology of generativity—biological, parental, technical, and cultural—which deepens understanding of how individuals contribute to future generations. While biological generativity aligns with procreation, the parental dimension centres on the transmission of values within family contexts. Technical generativity highlights the transfer of skills and systems of knowledge, distinguishing between performing tasks and teaching them. Cultural generativity, by contrast, concerns the preservation and evolution of traditions, symbols, and practices that sustain collective life. Kotre's (1984) typology underscores the multidimensional nature of generativity and its potential to extend far beyond the family unit, encompassing public, professional, and civic realms.

Taken together, these models underscore the importance of lifespan developmental frameworks that recognise the sustained potential for meaningful engagement in later life. Generativity, rather than diminishing with age, may be reframed as a vital resource for societal continuity and transformation, especially when approached through creative preservation lenses that emphasise moral commitment, relational responsibility, and cultural stewardship.

Cultural Generativity, Gero-Transcendence, and the Old–New

The concept of socio-cultural generativity—defined as "an other-regarding desire to invest one's substance in forms of life and work that will outlive the self" (Lang & Baltes, 1997, p. 27)—highlights a persistent concern with the transmission of meaning, values, and cultural continuity across generations. Erikson's later articulation of grand-generativity aligns with this perspective, framing the psychosocial task of late adulthood as a movement towards ego-transcendence—a developmental shift from self-centred achievement to the pursuit of enduring, intergenerational contributions. This perspective takes on particular educational and epistemological significance when situated within the broader framework of creative preservation (Verger et al., 2024; 2025).

Intergenerational interaction functions not merely as a form of social exchange but as a mode of cultural transmission and temporal pedagogy, fostering continuity across generations while cultivating the capacity to engage critically and creatively with inherited knowledge. In this sense, learning becomes an inter-temporal practice in which the pedagogical process entails the active reinterpretation and renewal of inherited forms rather than their displacement. Consistent with the framework of creative preservation, value is not derived from novelty or rupture with the past but from a heritage epistemology: a mode of knowing that re-explores the affordances of past objects and ideas to extend their longevity, resonance, and perceived value (Verger et al., 2024, p. 283).

Crucially, storytelling is a pivotal medium through which generative transmission is enacted. Drawing on Erikson's psychosocial theory and Kotre's (1984) elaboration of generativity, McAdams and Guo (2015) propose a processual model in which generativity is expressed through concrete actions, including generative narration. These narratives—intentional efforts to articulate the self through stories that foreground contributions to others—serve not only as personal recollections but also as moral and cultural templates, sustaining collective identity over time.

Through storytelling, older adults situate their experiences within broader historical, ethical, and communal frameworks. These stories often serve a pedagogical function, transmitting values that inform future decision-making and help prevent the repetition of past errors (Kotre, 1984). Cultural generativity is thus not a static trait but a dialogical, relational process shaped by individual identity and collective meaning-making. As McAdams and Guo (2015) caution, these autobiographical acts are not necessarily historically accurate but often assume a redemptive tone—shaped by prosocial motivations and intended to offer moral exemplars to subsequent generations. This emphasis on generative narration resonates further with Tornstam's (1996) theory of gero-transcendence, which suggests that ageing engenders a developmental shift from external accomplishments towards introspection and expanded temporal awareness. Here, "possibility thinking", as Glăveanu (2022) describes, is not only socially distributed but temporally stratified—bridging past, present, and future in imaginative and ethical engagement.

UNESCO (2003) characterises this generative contribution as intangible cultural heritage, encompassing a wide array of practices, expressions, knowledge, and skills—alongside the instruments, objects, artefacts, and cultural spaces associated with them—that communities, groups, and, in some cases, individuals recognise as integral to their cultural legacy. This form of living heritage reflects a dynamic interweaving of historical continuity, environmental context, and communal interaction, serving not only as a repository of identity but also as fertile ground for ongoing creativity. Kotre (1984) emphasises that generativity involves embedding influence in the fabric of future generations—through storytelling, mentorship, and cultural participation. From this perspective, generativity functions as a temporal bridge linking past experiences to future possibilities. Manheimer (1995) and Massie and Staude (2024) support this view, positioning older individuals as "living archives" who mediate between memory and emerging futures. The asymmetry between an expanding past and a shrinking personal future (Massie & Staude, 2024) may paradoxically amplify generative potential at the societal level.

Ultimately, these evolving conceptualisations of generativity, creativity, and cultural continuity challenge traditional deficit-based models of ageing, which foreground decline and dependency. Instead, they promote a vision of later life as a period marked by sustained agency, ethical responsibility, and ongoing contributions to the future. The embodiment of living heritage in older adults is not merely about preserving discrete sites or objects but about enacting dynamic processes through which those artefacts—and their associated memories and meanings—become relevant and transformative. As Auclair and Fairclough (2015, p. 4) observe, heritage includes "the various associations [objects] have for us, the memories they embody and the narratives they support", all of which "make us who we are, colour how we interact, and shape the futures that we can envisage and might create". In this sense, ageing becomes a generative force, replete with potential for cultural transmission, narrative continuity, and intergenerational engagement.

These insights are particularly pertinent in the broader context of contemporary global interconnectivity and environmental uncertainty, which amplify the ethical imperative to transmit ecological and cultural legacies. Laslett (2005, p. 123) contends that no generation can evade the responsibility of transmitting the world to those who follow. However, as Montuori and Donnelly (2020, p. 250) assert, future-making does not occur in abstraction; it arises from creative engagement with the present's material realities and inherited legacies. The future does not emerge *ex nihilo* but as an evolving synthesis of recombined pasts and responsive present actions.

Verger et al. (2025) argue that it is necessary to move beyond a production-based meritocracy towards a meritocracy of preservation—one that recognises the world as a "web of interdependencies to be respected, maintained, and stewarded" (p. 8). In this framework, value lies not solely in creating anew but in sustaining and reimagining what already exists. The authors further emphasise the "generative potential of existing relations to create new uses" (ibid., p. 8). This perspective is evident in the adaptive reuse of industrial sites across many European cities, where former warehouses—once symbols of economic decline—have been transformed into creative hubs. Rather than erasing the material and cultural traces of the past, architects and urban planners have integrated historical structures with contemporary design principles, sustainability measures, and digital technologies, thereby cultivating innovation through preservation (see also, Verger, Chap. 2, this volume).

Philosopher Vladimir Jankélévitch (2013) articulates this through his conception of progress as a cumulative process grounded in memory. He warns that without the anchoring presence of the past, futurition risks becoming empty innovation—mere novelty detached from historical substance. His notion of "slowed-down futurition" encourages a reflective, temporally layered approach to progress that integrates continuity and innovation. John Dewey (1957) similarly emphasises that habit, as the accumulation of adaptive past experiences, must remain responsive to emergent conditions. Habits should be critically evaluated and guided by reasonableness—defined as the capacity to perceive temporal continuities that connect actions to both past and future (ibid., p. 265). This ethical dimension situates human agency within a broader historical arc.

Donna Haraway's (2016) critique of autopoiesis and her metaphor of compost further reinforce this idea. In *Staying with the Trouble*, she proposes the notion of "sympoiesis"—a mode of collective becoming that emphasises co-creation across species, histories, and materialities (see also De Miranda, Chap. 3, this volume). Just as compost transforms decay into fertile soil, past experiences fertilise new growth. In this view, novelty is deeply entangled with memory, and meaningful futures arise from entangled rather than isolated origins. Similarly, Franklin-Phipps and Rath (in Kuby et al., 2018, p. 147) liken collaging to an epistemological practice—repurposing fragments of the past to construct new meanings. Rather than erasing history, collage reconfigures it, enabling alternative narratives that challenge dominant scripts. This practice exemplifies a speculative form of memory work, grounded in material and symbolic re-engagements with the past.

In her theorisation of culture, Margaret Mead (1970) delineates three cultural orientations: postfigurative, cofigurative, and prefigurative. Postfigurative cultures are those in which the values, norms, and behaviours of elders dominate, and traditions are rigidly preserved through established mores. Cofigurative cultures, while still acknowledging the authority of elders, place greater emphasis on peer influence and the role of contemporaries as behavioural models. Prefigurative cultures, by contrast, are marked by a decisive break from the past, in which younger generations imagine and shape new futures with diminished reliance on traditional norms.

Mead emphasises that the emergence of prefigurative cultures is not entirely independent of older generations, whom she later regarded as "living repositories of change" because they have witnessed drastic changes during their lifetimes (1974). For example, elders in Indigenous societies not only safeguard oral histories, ritual knowledge, and ecological practices but also represent the lived record of cultural adaptation to colonial encounters, environmental shifts, and social transformations. In this sense, they do not merely transmit continuity but exemplify how cultural knowledge itself is marked by processes of change, repositioning the elder as a carrier of both heritage and transformation. Mead contends that, to build a prefigurative culture "in which the past is instrumental rather than coercive", the very location of the future must be reconceptualised (Mead, 1970, p. 75). In this regard, she contrasts cultural perceptions of the future: in Western societies, it is typically envisioned as lying ahead; in Oceanic cultures, it is conceived as lying behind; and in Balinese culture, it is likened to "an underdeveloped film slowly developing". Thus, the spatial and temporal positioning of the future plays a critical role in shaping intergenerational dynamics and the cultural transmission of knowledge and values.

Future studies increasingly emphasise the need for historically informed foresight. Sandford (2019, p. 71) contrasts instrumental uses of the past—where information is abstracted for utilitarian ends—with more embedded approaches grounded in lived experience. For instance, policymakers draw selectively on historical data—such as records of agricultural yields or migration flows—abstracting them into models that serve immediate economic or political objectives. By contrast, an embedded approach is evident in the practices of farming communities that rely on generational memory of seasonal rhythms, soil conditions, and local ecological signals, knowledge inseparable from lived experience. Adam and Groves (2007) similarly call for future-making that integrates past, present, and future in socially constructive ways. Within this temporal ecology, heritage emerges as an interpretive and imaginative resource. Unlike history, which may be detached and linear, heritage is affective and embodied—allowing individuals and communities to reimagine the past as a means of constructing future trajectories (Sandford, 2019). Far from being static, heritage, as Holtorf and Bolin (2024) explain, is characterised by transformation and renewal. It supports human adaptability and creativity in the face of loss and uncertainty.

In this regard, the present moment is shaped by irreversible losses—of languages, customs, know-how, and ecological knowledge—contributing to the erosion of humanity's cultural and cognitive biodiversity and to the "standardisation of cultural identity into few stereotyped ways of life, ways of thinking and perceiving the world" (Lenzerini, 2011, p. 102). Haldon et al. (2020) argue that the adaptive strategies of past societies reveal not only their resilience but also their ability to reconfigure existing resources and practices in response to crisis—an orientation towards continuity through change that resonates strongly with the idea of creative preservation (Verger et al., 2024).

A particularly compelling example of *resilience through creative preservation* is provided by the Mental Health and Climate Change Alliance (MHCCA), which examines the restorative and adaptive capacities of community-led environmental initiatives. Through projects such as community gardens, seed banks, and fish hatcheries, the MHCCA shows how land-based practices can foster ecological regeneration and cultural continuity amid the disruptions of climate change. These initiatives not only secure traditional food sources threatened by environmental degradation but also engage younger generations in intergenerational knowledge exchange and cultural practice, thereby reinforcing community identity and collective resilience (Mental Health and Climate Change Alliance [MHCCA], 2025). This case demonstrates that traditional practices can simultaneously safeguard cultural heritage and generate renewed social and ecological value—embodying the principle of creative preservation, in which the past actively fertilises the emergence of the new.

Morewedge et al. (2005) note that our projections of the future are shaped by memory, albeit selectively and sometimes inaccurately. Still, engaging socially embedded memories remains essential for ethical foresight and sustainable planning. Heritage, in this sense, activates memory not merely to conserve the past but to transform it in service of the future. Within this intergenerational continuum, older adults emerge as pivotal figures—what Lang and Baltes (1997) call "keepers of meaning" (ibid., p. 27). As custodians of intangible cultural heritage, they mediate between continuity and transformation. For example, in the Basque Country, elder members of communities serve as custodians of intangible cultural heritage, actively preserving the Basque language (Euskara), traditional sports such as pelota, and oral storytelling practices, particularly the bertzolaritza, or improvised singing, promoted by the bertso-eskola movement as a form of intergenerational cultural transmission (Artetxe Sarsola, 2021). Simultaneously, they adapt these traditions to contemporary contexts, notably by promoting Euskara through digital media platforms, including those provided by the Basque public broadcaster EITB, as well as through community workshops and cultural festivals. Such initiatives ensure that these cultural practices remain both vibrant and socially relevant, fostering intergenerational transmission while responding to the dynamics of modern life.

Everyday heritage, as Auclair and Fairclough (2015, p. 4) remind us, is not a tourist commodity but a living practice—"the heritage of 'home,' not of 'holidays'". Its value lies in the social and cultural work of "doing heritage", a process of engaging with the past to navigate the present and create the future. Zittoun (2013) highlights the cognitive depth of older individuals, whose accumulated experiences enable them to reconstruct social realities and contribute meaningfully to future-making.

Kotre (1984) similarly frames generativity as a function of wisdom-in-action, a resource for cultural transmission. The concept of sustainability—often framed in forward-looking terms—gains substance when viewed through a historical lens. Rosen (2020) contends that sustainability should be conceived as a *historical science*—one that necessitates engagement with the broadest possible temporal scales to understand the continuity and transformation of human–environment relations. When framed through *the lens of creative preservation* (Verger et al., 2024; 2025), sustainability emphasises the persistence of older systems, practices, and ideas within the emergence of new ones, highlighting continuity rather than rupture as the foundation of innovation. The World Economic

Forum (2022) similarly asserts that ancient wisdom offers vital insights for addressing contemporary sustainability challenges. Practices such as traditional farming, water management, repairing, repurposing, and minimising waste—once integral to daily life—demonstrate how ancestral knowledge can inform resource-conscious and regenerative futures (Carrasco-Torrontegui et al. (2021).

The evolving discourse on socio-cultural generativity shows that older adulthood is not a denouement but a uniquely generative life stage. Through narrative, memory, habit, and heritage, older adults contribute to an ethical, imaginative, and ecologically grounded vision of the future—one in which continuity and innovation are not opposed but mutually sustaining.

Conclusion

In an age of uncertainty, it is increasingly urgent to recognise and value the multifaceted contributions of older people to the future. This chapter has argued that prevailing linear, deficit-based temporal frameworks constrain how ageing is imagined and experienced, producing reductive cultural scripts that relegate older adults to the past or cast them solely as relics of former significance. However, as explored throughout this discussion, such assumptions not only obscure the dynamic potential of older age but also undermine the very capacity for collective future-making.

Reconceptualising temporality as relational and non-linear enables us to disrupt dominant ageist ideologies and affirm the role of older individuals as critical agents in shaping ethically grounded, socially inclusive, and culturally rich futures. Contrary to prevailing stereotypes, this chapter has argued that older adults ought to be recognised as a vital generative force—not only for their creative capacities but also for their distinctive ability to mediate across temporal dimensions and foster intergenerational continuity. This provides practical cases and empirical grounding for the creative preservation framework (Verger et al., 2024; 2025).

Although age-related changes in cognitive processing speed and mental flexibility may pose constraints, older adults retain the capacity to acquire new competencies and engage adaptively with novel contexts. More significantly, the knowledge, values, and practices they transmit—though not always aligned with dominant conceptions of innovation—retain enduring salience and latent potential to shape future trajectories. Older people embody a living archive of memory, tradition, and learned adaptability.

Their experiences are not merely historical residues but interpretive frameworks that inform present decision-making and foster resilience in the face of contemporary crises. As Massie and Staude (2024, p. 583) observe, "the temporality of ageing not only intertwines past, present, and future but also biological, personal, and social time", thereby grounding emergent possibilities while enabling development and progression.

At the heart of this relational temporality lies the practice of creative preservation—a generative process through which older adults sustain, reinterpret, and reimagine cultural memory as a living, evolving inheritance (Verger et al., 2024; 2025). Creative preservation transcends the conventional dichotomy between innovation and conservation: it is not the passive safeguarding of the past but the imaginative renewal of what is worth carrying forward. It is a form of cultural care that insists on the vitality of memory, treating preservation as a creative act that continually reopens the archive to reinterpretation and transformation. In this sense, creative preservation serves as a counterpoint to the rapid obsolescence and temporal acceleration characteristic of late modernity, offering instead a slower, more reflective rhythm of cultural continuity.

This chapter has examined the interrelated concepts of socio-cultural and cultural generativity, alongside the notion of slowed-down futurition, in which future thinking unfolds in a more measured and reflective manner. Through acts of creativity, storytelling, mentorship, and cultural transmission, older adults may function as agents of "slow futurition"—a contextually grounded and ethically attuned mode of future-making that resists commodified acceleration and fosters more participatory, inclusive, and less instrumentalised engagements with the possible. Ultimately, older adults enact a form of creative preservation that constitutes a distinctive mode of temporal stewardship.

Conceived as embodiments of living heritage, older individuals offer a crucial orientation for interpreting the present—an orientation "without which the future would be rudderless" (Bradfield et al., 2016, p. 570). This reframing presents ageing not as a retreat from possibility but as a temporal horizon uniquely suited to bridging continuity and transformation. Through creative preservation (Verger et al., 2024; 2025), older adults become both curators and innovators of cultural meaning, ensuring that collective memory does not calcify into nostalgia but remains porous, dialogical, and responsive to emergent realities. Their preservation work is thus profoundly creative: it entails discerning what must endure, what must evolve, and what must be relinquished.

Moreover, the embodied knowledge and imaginative capacities of older adults can serve as powerful counterpoints to the disorientation caused by technological and ecological upheaval. As custodians of intangible cultural heritage and ethical memory, they offer vital resources for navigating complexity, transmitting values, and grounding innovation in historically conscious practices. Their contributions extend beyond personal wisdom or private reflection; they include the active curation of future heritages—sociocultural practices, ethical commitments, and creative expressions that nourish intergenerational dialogue and sustainable world-building. In this sense, creative preservation is not merely a descriptive concept but a normative imperative: a practice that secures the depth, coherence, and ethical texture of the futures we seek to build (Verger et al., 2024; 2025).

By embracing a temporally complex, non-ageist framework, we can foster a more inclusive, pluralistic vision of the future—one that honours the full lifespan of human development and reclaims later life as a period of contribution, relevance, and generativity. Far from being obsolete, older adults are indispensable to the imaginative, ethical, and cultural scaffolding of the futures we hope to inhabit. Their participation is not only desirable—it is necessary. While the impermanence that permeates contemporary life may be inescapable (Toffler, 2022), forms of transcendence—particularly ego-transcendence, as embodied by older people—may help counterbalance the pervasive sense of transience that characterises the human condition. The future, to be truly sustainable and meaningful, must be co-authored by those who carry its longest memories and remain invested in its most enduring possibilities. In this co-authorship, creative preservation (Verger et al., 2024) emerges as the connective art of futurity itself—a practice through which the wisdom of the past is not only remembered but continually reimagined to sustain the creative conditions of collective life.

References

Adam, B., & Groves, C. (2007). *Future matters: Action, knowledge, ethics* (Supplements to the Study of Time, Vol. 3). Brill. https://doi.org/10.1163/ej.9789004161771.i-218

Aristotle. (1936). *Problems* (W. S. Hett, Trans.). Harvard University Press. (Loeb Classical Library No. 316).

Aristotle. (1985). On the soul (De Anima) (J. A. Smith, Trans.). In J. Barnes (Ed.), *The complete works of Aristotle, volume one: The revised Oxford translation* (pp. 641–692). Princeton University Press. https://doi.org/10.1515/9781400835843-015

Artetxe Sarasola, M. (2021). Bertso eskola, youth identities and new speakers. *Journal of Multilingual and Multicultural Development, 43*(1), 43–54. https://doi.org/10.1080/01434632.2021.1988620

Auclair, E., & Fairclough, G. (Eds.). (2015). *Theory and practice in heritage and sustainability: Between past and future* (1st ed.). https://doi.org/10.4324/9781315771618

Ayalon, L., & Tesch-Römer, C. (2018). Introduction to the section: Ageism—Concept and origins. In *Contemporary perspectives on ageism* (pp. 1–11). Springer Open.

Baltes, P. B., Reese, H. W., & Lipsitt, L. P. (1980). Life-span developmental psychology. *Annual Review of Psychology, 31*, 65–110. https://doi.org/10.1146/annurev.ps.31.020180.000433

Beard J.R., Officer A., Araujo de Carvalho I., Sadana R. , Margriet Pot A., Michel J., Lloyd-Sherlock P. , E Epping-Jordan J., Peeters G., Retno Mahanani W., Amuthavalli Thiyagarajan J., Chatterji S.(2016) *The World report on ageing and health: a policy framework for healthy ageing*, The Lancet, Volume 387, Issue 10033, Pages 2145-2154, ISSN 0140–6736, https://doi.org/10.1016/S0140-6736(15)00516-4.

Bergson, H. (1910). *Time and free will: An essay on the immediate data of consciousness* (F. L. Pogson, Trans.). George Allen & Unwin.

Bradfield, R., Derbyshire, J., & Wright, G. (2016). The critical role of history in scenario thinking: Augmenting causal analysis within the intuitive logics' scenario development methodology. *Futures, 77*, 56–66. https://doi.org/10.1016/j.futures.2016.02.002

Brito, D. V. C., Esteves, F., Rajado, A. T., Silva, N., ALFA Score Consortium, Araújo, I., Bragança, J., Castelo-Branco, P., & Nóbrega, C. (2023). Assessing cognitive decline in the ageing brain: Lessons from rodent and human studies. *npj Ageing, 9*, 23. https://doi.org/10.1038/s41514-023-00120-6

Butler, R. N. (1969). Age-ism: Another form of bigotry. *The Gerontologist, 9*, 243–246.

Camic, P. M., Crutch, S. J., Murphy, C., Firth, N. C., Harding, E., Harrison, C. R., Howard, S., Strohmaier, S., Van Leeuwen, J., West, J., Windle, G., Wray, S., & Zeilig, H. (2018). Conceptualising and understanding artistic creativity in the dementias: Interdisciplinary approaches to research and practise. *Frontiers in Psychology, 9*, 1842. https://doi.org/10.3389/fpsyg.2018.01842

Carrasco-Torrontegui, A., Gallegos-Riofrio, C., Delgado-Espinoza, F., & Swanson, M. (2021). Climate change, food sovereignty, and ancestral farming technologies in the Andes. *Current Developments in Nutrition, 5*(Suppl. 4), 54–60.

Cicero. (1923). *Cato Maior de Senectute* (N. A., Trans.). Retrieved March 15, 2025, from https://penelope.uchicago.edu/Thayer/E/Roman/Texts/Cicero/Cato_Maior_de_Senectute/text*.html

Costa, D. L. (1998). *The evolution of retirement: An American economic history, 1880–1990*. University of Chicago Press.

Craft, A. (2000). *Creativity across the primary curriculum: Framing and developing practice*. Routledge.

Craft, A. (2001a). *An analysis of research and literature on creativity in education*. Report prepared for the Qualifications and Curriculum Authority. http://www.creativetallis.com/uploads/2/2/8/7/2287089/creativity_in_education_report.pdf

Craft, A. (2001b). Little c creativity. In A. Craft, B. Jeffrey, & M. Leibling (Eds.), *Creativity in education* (pp. 45–61). Continuum.

Craft, A. (2003). The limits to creativity in education: Dilemmas for the educator. *British Journal of Educational Studies, 51*(2), 113–127.

Datta Roy, M., & Panda, M. (2022). Is generativity Erikson's answer to human ageing in the neoliberal world? *Journal of Ageing Studies, 62*, 101057. https://doi.org/10.1016/j.jageing.2022.101057

Dewey, J. (1957). *Human nature and conduct: An introduction to social psychology*. Modern Library.

Edmonson, R. (2009). *Valuing older people*. Bristol University Press.

Erikson, E. H. (1950). *Childhood and society*. W. W. Norton.

Erikson, E. H. (1959). *Identity and the life cycle*. International Universities Press.

Erikson, E. H. (1982). *The life cycle completed*. Norton.

Eshel, A. (2012). *Futurity: Contemporary literature and the quest for the past*. University of Chicago Press.

Facer, K. (2023). Possibility and the temporal imagination. *Possibility Studies & Society, 1*(1–2), 60–66. https://doi.org/10.1177/27538699231171797

Facer, K. (2024, July). *Possibility and the temporal imagination* [Keynote]. Possibility Studies Network Conference, Cambridge.

Fried LP, Freedman M, Endres TE, Wasik B.(1997) Building communities that promote successful aging. West J Med. 1997 Oct; 167(4): 216–9.

Glăveanu, V. P. (2022). Possibility studies: A manifesto. *Possibility Studies & Society, 1*, 1–6. https://doi.org/10.1177/27538699221127580

Gordon, J. K., & Jordan, L. M. (2017). Older is wiser? It depends who you ask … and how you ask. *Ageing, Neuropsychology, and Cognition, 24*(1), 94–114. https://doi.org/10.1080/13825585.2016.1171292

Gubrium, J. F. (1995). Voice, context, and narrative in aging research. *Canadian Journal on Aging, 14*(1, Suppl 1), 68–81. https://doi.org/10.1017/S0714980800005432

Guerlac, S. (2006). *Thinking in time: An introduction to Henri Bergson*. Cornell University Press.

Gullette, M. M. (2018). Against 'ageing'—How to talk about growing older. *Theory, Culture & Society, 35*(7–8), 251–270. https://doi.org/10.1177/0263276418811034

Hickson, J., & Housley, W. (1997). Creativity in later life. *Educational Gerontology, 23*(6), 539–547.

Haldon J, Eisenberg M, Mordechai L, Izdebski A, White S. Lessons from the past, policies for the future: resilience and sustainability in past crises. Environ Syst Decis. 2020; 40(2): 287–297. https://doi.org/10.1007/s10669-020-09778-9

Haraway, D. J. (2016). *Staying with the Trouble: Making Kin in the Chthulucene.* Durham: Duke University Press

Holtorf, C., & Bolin, A. (2024). Heritage futures: A conversation. *Journal of Cultural Heritage Management and Sustainable Development, 14*(2), 252–265. https://doi.org/10.1108/JCHMSD-09-2021-0156

Jankélévitch, V. (2013). *Forgiveness.* University of Chicago Press.

Jolanki, O. U. (2009). Agency in talk about old age and health. *Journal of Ageing Studies, 23*(4), 215–226.

Jonson, H. (2012). We will be different! Ageism and the temporal construction of old age. *The Gerontologist, 53*(2), 198–204.

Kaufman, J. C., & Beghetto, R. A. (2009). Beyond Big and Little: The Four C Model of Creativity. *Review of General Psychology, 13*(1), 1–12.

Kemper, S., & Harden, T. (1999). Experimentally disentangling what's beneficial about elderspeak from what's not. *Psychology and Ageing, 14*(4), 656–666. https://doi.org/10.1037/0882-7974.14.4.656

Khammash, D., Rajagopal, S., & Polk, T. (2023). The neurobiology of aging. In M. Zigmond, C. Wiley, & M. Chesselet (Eds.), *Neurobiology of brain disorders* (pp. 977–993). Academic Press.

Kontos P, Martin W. (2013) Embodiment and dementia: exploring critical narratives of selfhood, surveillance, and dementia care. Dementia (London). May; 12(3):288–302. https://doi.org/10.1177/1471301213479787

Kotre, J. (1984). *Outliving the self: Generativity and the interpretation of lives.* Johns Hopkins University Press.

Kuby, C., Spector, K., & Johnson Thiel, J. (Eds.). (2018). Posthumanism and Literacy Education: Knowing/Becoming/Doing Literacies (1st ed.). Routledge. https://doi.org/10.4324/9781315106083

Lang, F. R., & Baltes, M. M. (1997). Being with people and being alone in late life: Costs and benefits for everyday functioning. *International Journal of Behavioral Development, 21*(4), 729–746. https://doi.org/10.1080/016502597384640

Laslett, P. (2005). Interconnections over time. *Journal of Classical Sociology, 5*(2), 123–233.

Lazar, N. C. (2019). *Out of joint: Power, crisis, and the rhetoric of time.* Yale University Press. https://doi.org/10.12987/9780300245172

Lenzerini, F. (2011). Intangible cultural heritage: The living culture of peoples. *European Journal of International Law, 22*(1), 101–120. https://doi.org/10.1093/ejil/chr006

Levy, B. (2022). *Breaking the age code: How your beliefs about ageing determine how long and well you live.* William Morrow.

Levy, B. R. (2009). Stereotype embodiment: A psychosocial approach to ageing. *Current Directions in Psychological Science, 18*(6), 332–336. https://doi.org/10.1111/j.1467-8721.2009.01662.x

Löckenhoff, C. E., De Fruyt, F., Terracciano, A., McCrae, R. R., De Bolle, M., Costa, P. T., Jr., Aguilar-Vafaie, M. E., Ahn, C.-k., Ahn, H.-n., Alcalay, L., Allik, J., Avdeyeva, T. V., Barbaranelli, C., Benet-Martínez, V., Blatný, M., Bratko, D., Cain, T. R., Crawford, J. T., Lima, M. P., et al. (2009). Perceptions of ageing across 26 cultures and their culture-level associates. *Psychology and Ageing, 24*(4), 941–954. https://doi.org/10.1037/a0016901

Löckenhoff, C. E., Drewelies, J., Duezel, S., Steinhagen-Thiessen, E., Demuth, I., Freund, A. M., ... & Gerstorf, D. (2022). Sociohistorical change in urban older adults' perceived speed of time and time pressure. *The Journals of Gerontology: Series B 77*, (3), 457–466.

Luppi, E. (2015). Training to intergenerational dialogue and solidarity. *Jeugbeleid, 9*, 157–164.

Machado de Oliveira, V. (2021). *Hospicing Modernity: Facing Humanity's Wrongs and the Implications for Social Activism.* London: Penguin Random House.

Malendowicz, P. (2019). The concept of "the return to the past" as an inspiration for the anti-civilization project of utopian primitivist thought. *Studia Politonogicze, 53*, 200–214.

Manheimer, R. J. (1995). Redeeming the ageing self: John Kotre, George Drury, and cultural generativity. *Journal of Ageing Studies, 9*(1), 13–20. https://doi.org/10.1016/0890-4065(95)90022-5

Massie, P., & Staude, M. (2024). Ageing in the world. *Continental Philosophy Review, 57*, 565–584. https://doi.org/10.1007/s11007-024-09630-w

McAdams, D. P., & Guo, J. (2015). Narrating the generative life. *Psychological Science, 26*, 475–483.

McCrae, R. R., Arenberg, D., & Costa, P. T., Jr. (1987). Declines in divergent thinking with age: Cross-sectional, longitudinal, and cross-sequential analyses. *Psychology and Ageing, 2*(2), 130–137. https://doi.org/10.1037/0882-7974.2.2.130

Mead, M. (1970). *Culture and commitment: A study of the generation gap.* Natural History Press.

Mead, M. (1974, December). Grandparents as educators. *Teachers College Record, 76*, 240–245.

MHCCA. (2025). *Chapter 4: Land-based healing and cultural resilience in healing with the land.* Resources for Indigenous Communities. https://mhcca.ca/

indigenous-communities/chapter-4-land-based-healing-and-cultural-resilience#:~:text=Cultural%20resilience%20refers%20to%20a,pain%20and%20to%20encourage%20renewal

Montuori, A., & Donnelly, G. (2020). Creativity and the future. In M. Runco & S. Pritzker (Eds.), *Encyclopedia of creativity* (Vol. 1, 3rd ed., pp. 250–257). Academic Press.

Morewedge, C. K., Gilbert, D. T., & Wilson, T. D. (2005). The least likely of times: How remembering the past biases forecasts of the future. *Psychological Science, 16*(8), 626–630.

Moscovici, S. (1988). Notes towards a description of social representations. *Journal of European Social Psychology, 18*, 211–250.

Mullarkey, J. (2007). Life, Movement and the Fabulation of the Event. *Theory, Culture & Society, 24*(6), 53–70.

Murman, D. L. (2015). The impact of age on cognition. *Seminars in Hearing, 36*(3), 111–121. https://doi.org/10.1055/s-0035-1555115

Nadobnik, N., Durczak, K., & Ławrynowicz-Ross, M. (2021). Temporality and the ageing self: How subjective time is folding over its linear progression. *Journal of Ageing Studies, 57*, 9–12.

Palmiero, M., Di Giacomo, D. & Passafiume, D. Creativity and dementia: a review. *Cogn Process 13*, 193–209 (2012). https://doi.org/10.1007/s10339-012-0439-y

Rosen, I. (2020). Sustainability as a historical science: A focus on environment and human culture. *The Post Hole*. https://www.theposthole.org/read/article/485

Ross, M., Lachmann, T., Jaarsveld, S., Riedel-Heller, S. G., & Rodriguez, F. S. (2023). Creativity across the lifespan: Changes with age and with dementia. *BMC Geriatrics, 23*, 307. https://doi.org/10.1186/s12877-023-03964-5

Rudnicka E, Napierała P, Podfigurna A, Męczekalski B, Smolarczyk R, Grymowicz M. The World Health Organization (WHO) approach to healthy ageing. Maturitas. 2020 Sep;139: 6–11. https://doi.org/10.1016/j.maturitas.2020.05.018

Rugh, M. M. (1991). Creativity and life review in the visual arts: The transformative experience of Florence Kleinsteiber. *Generations: Journal of the American Society on Ageing, 15*(2), 27–31.

Sandford, R. (2019), Thinking with heritage: Past and present in lived futures, Futures, Volume 111, Pages 71–80,

Singer, P. (1981). *Expanding the circle*. Princeton University Press.

Strawson, G. (2004). Against narrativity. *Ratio, 17*(4), 428–452. https://doi.org/10.1111/j.1467-9329.2004.00264.x

Thomas, C. (1999). *Female forms: Experiencing and understanding disability*. Open University Press.

Thönes, S., & Stocker, K. (2019). A standard conceptual framework for the study of subjective time. *Consciousness and Cognition, 71*, 114–122.

Thornton, J. E. (2002). MYTHS OF AGING OR AGEIST STEREOTYPES. *Educational Gerontology, 28*(4), 301–312. https://doi.org/10.1080/036012702753590415

Toffler, A. (2022). *Future shock*. Ballantine Books.

Tornstam, L. (1996). Caring for the elderly: Introducing the theory of Gerotranscendence as a supplementary frame of reference for caring for the elderly. *Scandinavian Journal of Caring Sciences, 10*, 144–150.

Tromp, C., & Glăveanu, V. P. (2023). A model of creative ageing (MOCA): Unlocking the potential of constraints for creativity in older adults. *Psychology of Aesthetics, Creativity, and the Arts*. Advance online publication. https://doi.org/10.1037/aca0000538

UNESCO. (2003). *Convention for the safeguarding of the intangible cultural heritage*. https://ich.unesco.org/en/convention

Verger, N. B., Duymedijan, R., & Glăveanu, V. P. (2025). The meritocracy of preservation: Reimagining merit beyond production. *Organization, 1–19*. https://doi.org/10.1177/13505084251347519

Verger, N. B., Duymedijan, R., Wegener, C., & Glăveanu, V. P. (2024). Creative preservation: A framework of creativity in support of degrowth. *Review of General Psychology, 28*(3), 268–281. https://doi.org/10.1177/10892680241256312

Veríssimo, J., Verhaeghen, P., Goldman, N., Weinstein, M., & Ullman, M. T. (2022). Evidence that ageing yields improvements as well as declines across attention and executive functions. *Nature Human Behaviour, 6*, 97–110. https://doi.org/10.1038/s41562-021-01169-7

Walker, A., & Maltby, T. (2012). Active ageing: A strategic policy solution to demographic ageing in the European Union. *International Journal of Social Welfare, 21*(Suppl. 1), S117–S130. https://doi.org/10.1111/j.1468-2397.2012.00871.x

Weick, K. E. (1979). *The social psychology of organizing* (2nd ed.). Addison Wesley.

Weick, K. E., Sutcliffe, K. M., & Obstfeld, D. (2005). Organizing and the process of sensemaking. *Organization Science, 16*(4), 409–421. https://doi.org/10.1287/orsc.1050.0133

Whiting, W. L., IV, & Smith, A. D. (1997). Differential age-related processing limitations in recall and recognition tasks. *Psychology and Ageing, 12*(2), 216–224. https://doi.org/10.1037/0882-7974.12.2.216

World Economic Forum. (2022, March). *What can ancient wisdom teach us about sustainability?* https://www.weforum.org/stories/2022/03/ancient-wisdom-can-teach-us-a-lot-about-sustainability/

Zaninotto P, Batty GD, Allerhand M, Deary IJ. Cognitive function trajectories and their determinants in older people: 8 years of follow-up in the English

Longitudinal Study of Ageing. J Epidemiol Community Health. 2018 Aug; 72(8): 685–694. https://doi.org/10.1136/jech-2017-210116

Zittoun, T. (2023). *Toward a dialogical understanding of objects in subjective experience and social practices.*

Zimbardo, P., & Boyd, J. (2008). *The time paradox: The new psychology of time that will change your life.* Free Press.

Zittoun, T., Valsiner, J., Vedeler, D., Salgado, J., Gonçalves, M., & Ferring, D. (2013). *Human development in the lifecourse.* Melodies of living. Cambridge: Cambridge University Press.

Zittoun, T. (2022). Old age. In *The Palgrave Encyclopedia of the Possible* (pp. 937–944). Palgrave Macmillan

CHAPTER 5

Expanding Possibilities for Research Methodologies that Inspire Change: Why "Thing-Power" Matters

Pamela Burnard, Laura Colucci-Gray, Donald Gray, and Tim Palmer

P. Burnard (✉)
Faculty of Education, University of Cambridge, Cambridge, UK
e-mail: pab61@cam.ac.uk

L. Colucci-Gray
Moray House School of Education and Sport, University of Edinburgh, Edinburgh, UK
e-mail: laura.colucci-gray@ed.ac.uk

D. Gray
School of Education, Aberdeen University, Aberdeen, UK
e-mail: d.s.gray@abdn.ac.uk

T. Palmer
Guildhall School of Music & Drama, London, UK
e-mail: t.palmer@trinitylaban.ac.uk

N. B. Verger, D. Mattson (eds.), *Possibility Studies and Societal Futures*, Palgrave Advances in Possibility Studies,
https://doi.org/10.1007/978-3-032-26379-7_5

Introducing Posthumanist and New Materialist Framing Possibilities

We live at a time of profound social and ecological transformations. The challenge of complex, intractable problems has highlighted the limits of a world-view fragmented into disciplines, calling instead for transdisciplinary approaches that are open to a multiplicity of partial but equally legitimate perspectives (Funtowicz & Ravetz, 1994; Saltelli, 2024). Moving beyond the empirical and reductionist focus that has characterised modern Western science and adopting more relational forms of inquiry, such as dialogue and engagement with diverse forms of knowing, may lead to new questions, spark unexpected curiosities, and open up possibilities for a renewed role for researchers in society (Meskens, 2024). In other words, we need to research the limits of our knowing and the possibility of knowing again (Glăveanu, 2023). We also need to research where new responses to current problems may come from, given a different understanding of how and why methodology matters (Lury, 2020). We need methodologies that advance the situatedness of knowledges (Frank et al., 2024) and practices of creative preservation that can prompt reflexivity about how materials and creative acts can contribute to desirable collective futures (Verger et al., 2024; 2025).

This point has been well argued by Rosi Braidotti (2019, p. 11):

> The primary task of posthuman critical thought is to track and analyse the shifting grounds on which new, diverse and even contradictory understandings of the human are currently being generated, from a variety of sources and cultures and traditions … it means adopting an internally differentiated and grounded notion of being human … Recognising the embodied and embedded, relational and affective positions of humans is a form of situated knowledge that enhances the singular and collective capacity for both ethical accountability and alternative ways of producing knowledge.

This chapter examines how specific instances of relational-material practice influence the enactment of research methodologies in education. By embracing relational and material considerations, we also offer a new approach to integration across disciplines, one that is deeply entangled with places, bodies, and affect that delineate what matters in research.

As Braidotti (2019) explains, posthuman practices invite "a materialist process ontology" (ibid., p. 53), shifting the focus away from humans as the sole producers of knowledge. She emphasises an "assemblage within a common life-space that the subject never masters nor possesses but merely inhabits, crosses, always in a community, a pack, a group, or a cluster" (ibid., p. 193).

In posthumanist and new materialist inquiry, the complex dynamics of relational assemblages are thus mapped from within, rather than reduced to easily digestible chunks of information from the outside (Fox & Alldred, 2017), as is often the case in conventional research enquiry. For example, alongside the typical focus on objectification and measurement within specific boundaries, research practices may also include first-person approaches, through which the researcher enters into dialogue with the complexity of the context, the shifting of boundaries, and the plurality of subjects involved.

As Braidotti (2019) explains, "within" entails the capacity to challenge the separation of knowledges, tools, and bodies in "disciplinary bounded" spaces. The bodies of non-human and more-than-human entities and their mutual boundaries are continually renegotiated through intra-action, that is, processes of mutual co-constitution or assemblages of complex, evolving relations (Barad, 2007). Thus, in this chapter, we employ a new materialist framing of materiality that recognises the active role non-human matter plays in shaping, producing, and transforming our worlds and experiences, thereby challenging the idea of human exceptionalism (Braidotti, 2019) and re-seeing research as relational assemblages.

Working with the Vitality of Materials

In this chapter, we argue that the form of the assemblages in our research depends on how material things combine, interconnect, and intra-act to give rise to an essentially discursive, material, and expressive entity. To do this, we elaborate on Deleuze and Guattari's (2013) concept of an "assemblage", which, rather than a simple composition of elements, is a dynamic, tri-dimensional arrangement between two (or more) semi-autonomous formations. In Jane Bennett's (2010, p. 35) view, "an assemblage owes its agentic capacity to the vitality of the materials that constitute it".

Specifically, Bennett's concept of "thing-power…gestures towards the strange ability of ordinary, man-made items to exceed their status as objects and…manifest traces of independence or aliveness, constituting the outside of our own experience" (ibid., p. xvi). People tend to think of objects (e.g. a piece of soil, a musical instrument, a garden tool) as inert things that just sit there until a person picks them up and does something with them. Jane Bennett's concept of "thing-power" challenges that assumption. In her book Vibrant Matter, she argues that materials and objects aren't passive. They push back, shape what we do, and influence outcomes in ways we don't always notice. A violin's grain of wood affects the sound a musician can produce. Soil in a school garden responds differently depending on weather and season, redirecting what children learn. The objects aren't just being used. Instead, we could say they're participating. Bennett's point isn't that things have intentions the way people do, but that they carry a vitality and force of their own that matters. Bennett's term encourages a shift from viewing the world as a collection of dead resources to seeing it as a vibrant, interconnected web of forces.

Thingness challenges the frameworks through which we traditionally research a specific object, act, thought, or discipline, and the frameworks through which we presume to know. By attending to the "thingness" of bodies and matter, we are provoked to see beyond linear spatio-temporalities and dualisms conventionally assigned to the problematic of human-centred theories. Thingness refocuses our attention, even prompting a reimagining of ourselves as researchers, and challenges anthropocentric perspectives on reality and what counts as data by attending to entanglements of relations and a re-attunement with human, non-human, and more-than-human worlds.

But why "thing-power" and why "assemblages"? What new possibilities do these threshold concepts open up? The necessary question here—as with all research—is: what does it allow us to do in the world?

Understanding the deep structures that shape how humans relate to one another, the planet, and other life forms is a vital and urgent step towards doing things differently. Zakiyya Iman Jackson, in her 2020 book *Becoming human*, invites us to radically shift human–non-human binaries so that we can navigate the world's multiple complexities and think differently about matter, materiality, and posthumanism. She inspires us to generate alternative frameworks of thought and relationality that not only disrupt the racialisation of the human–animal distinction found in Western science and philosophy but also challenge the epistemic and material terms through which the spectre of animal life acquires its authority. She also invites serious rethinking of how to enact and expand methodological possibilities for change.

In this collaboration, we map "thing-power" enactments across diverse research terrains, featuring: (i) schoolchildren's agentic entanglements in

STEAM gardening practices; (ii) secondary school students' mathematical art practices; and (iii) professional music performance practices. We show how these discursive-material practices materially evidence Jane Bennett's (2010) notion of thing-power. The emphasis is on the embodied and situated nature of knowledges, with close attention to multimodal communication and a redefinition of research territories through the horizontal, methodological practice of the "as" across each study. Secondly, we ask what it means, methodologically, to (re)-(con)figure education as possibilities for uncertain futures. We ask: how do "thing-power" and "thingness" generate new ways of reconfiguring methodologies for inquiry?

From here, we craft a cartographic assemblage of research practices to reimagine and reconfigure education through posthumanist possibilities that engage with the urgencies of our time.

Evidencing the Materiality of "Thing-Power" Through STEAM Gardening

The 68th UN General Assembly declared 2015 the International Year of Soils. The declaration came from the longstanding recognition of soil's undervaluation and its role in planetary functions and long-term survival. Soil has been central to human existence since our origins, not only as the immediate source of food but also in the regulation of microclimates, adapting its response to and with human practices. Yet, ten years later, the state of soils worldwide has further deteriorated, and soil is now recognised as one of the critically endangered planetary boundaries, central to the preservation of biodiversity as well as climate regulation (Giuffredi et al., 2025).

From a posthumanist perspective, enquiring into soil is not only an inquiry into the state of soil itself; it is also about the relationship between humans and the planet. This recognition is even more important in education, which, it has been argued, has become increasingly removed from ecological principles and the relational matrix that supports life (Blenkinsop & Kuchta, 2024). However, soil, as an entanglement of different elements, does not sit comfortably within formal educational structures; it pushes at the edges of curricular partitions, transgresses formal and informal spaces, and challenges the academic–vocational divide. Drawing together experiences from a series of projects that engaged us directly, we set out to inquire into educational practices that were generative of new understandings and possibilities for thinking-with and making-with soil in formal education, beyond the service of a logic of performativity and perpetual economic growth (Gray et al., 2019; Hancock et al., 2025).

A common thread across all these projects is that soil comes to life as an entanglement of elements and intra-acting forces (Barad, 2007). Transgressing disciplinary boundaries and coming together as a nexus of knowledge, affect, and embodied practices, these soil-based projects incorporated a multiplicity of perspectives, both disciplinary and experiential, as different ways of knowing, spanning formalised models as well as local and indigenous practices (Scoones, 2015). In considering possibilities for educational futures, we are thus invited to think differently about soil: as a subject in a joint inquiry, materially and physically entangled with the materialities of human and more-than-human lives, rather than an object of analysis at a distance. In this view, the "thing-power" of soil makes itself felt through the generative force of co-constitutive agencies, embodied and embedded in the capacity of all members to both affect and be affected (Haraway, 2016).

To bring forth such power, we draw on the acronym STEAM (traditionally Science, Technology, Engineering, and Mathematics with the arts) to introduce negative dialectics. On the one hand, we acknowledge and push against the "extractive mindset" that has taken hold of educational systems driven by measurement and the model of economic growth underpinning STEM education, even as the arts are brought into the mix (Colucci-Gray, 2022). In this view, the acronym emphasises the singular disciplines, each with its own body of specialist knowledge, languages, and units of analysis, all deemed to be transferred to students. On the other hand, we draw upon the synergistic possibilities of letting arts and sciences "teach" together, that is, in the manner of science as art and art as science (Burnard et al., 2021).

While the sciences may be concerned with evidence and effects, the arts may be pitched towards the future and towards making sense of the non-observable, but both engage with artefacts as products of human construction (Eisner, 2003) and are "involved in a kind of composition or in the 'making' of something in the here and now, that also gestures elsewhere" (O'Sullivan, 2024, p. 128). Arguably, this approach is essential for addressing the complexity of a world in transformation, in which no single perspective is valid at all times, and a plurality of legitimate perspectives is needed (Funtowicz & Ravetz, 1994).

But it goes further. Following O'Sullivan (2024), the preposition "as" is the linguistic device that engages and embodies fictioning practices by drawing upon a familiar reality in the past, while conjuring new imaginations in the present-future. This approach, drawn from the arts, also resonates with the concept of "creative preservation", which refers to practices of creation that re-value materials by updating and adapting them, or re-expressing them in another way, through the exploration of their affordances (Verger et al., 2024, p. 270).

In our work, the transdisciplinary concept of a STEAM garden is a materially discursive figuration that troubles disciplinary fragmentation and pushes at the edges of the possible, "pre-figuring" what has not yet been thought or materialised in education (Colucci-Gray, 2025; Gray et al., 2019). Within this frame, the experiences we have been involved in over the long arch of time spanning 2013 through to 2025 are diffracted through the lens of STEAM gardens, both to document them through evidence, in the manner of the sciences, and to make sense of the experiences that arose, as in the practice of an artist. We explore the surfacing patterns of relations and the moments of intense affectivity in which the relations between humans and soils are re-viewed and re-materialised.

Specifically, we draw on a selection of projects in which we were directly involved as part of research collaborations and professional development courses with secondary school teachers in Argentina and Scotland (2013–2016), and with primary and secondary school students in Scotland (2017 onwards). Following Fox and Alldred (2017), in each project we engaged with the thing-power of soil as the diffractive apparatus that enabled the juxtaposition of the view of the parts with the relational figurations of entanglements; the felt patterns of intra-actions (not just inter-actions) and the patterns of influence, drawing together multiple perspectives and incorporating our own creative insights as researchers engaged in the process, as a means to specify a particular data analysis "cut" (Taguchi & Palmer, 2013).

Criss-Crossing with Soil

The first project, which ran from 2013 to 2016, was a collaboration with the University of San Andrés in Buenos Aires, Argentina, and focused on creativity in science education. In the second year, we worked with a group of secondary school science teachers in Scotland, linking to the 2015 UN International Year of Soils. Given the challenge of addressing such themes within the subject-bound secondary school curriculum, we drew on the wider international call for soil preservation to introduce teachers to different modalities of inquiry that drew on the idea of relational assemblages through the reinterpretation and re-seeing of materials across different media.

Participants were first invited to consider their own understanding of soil, drawing on their different sets of knowledge and a selection of interdisciplinary literature; then to draw how they perceived it and how the topic was presented in the curriculum. During these activities, they shared ideas with others in the group and engaged in sensory and artistic explorations of soil, suggesting metaphors and drawing on them as a stimulus for free writing. Following the free writing, we asked participants to extract phrases and ideas they agreed on and to write these on strips of paper, ready to be reassembled in a poetic form. These idea poems were then performed for the class in a variety of ways (Fig. 5.1 a–c).

Thing-Power of Soil and the Embodied Practice of Writing

This first embodied experience of the thing-power of soil evolved into the following embodied practice of writing, as the artefacts that participants produced (Fig. 5.1b) were not merely recollections of what happened or what they read. Following Bennett (2010), when seen together and through one another across the three pictures, the thing-power of half-drunk coffee cups, biscuit crumbs, bits of paper, and a laid-down pencil emerges in an assemblage that is transient and "vibratory": "in the mood or style of an open whole in which the membership changes over time, and the members themselves undergo internal alteration" (ibid., p. 35). Seeing knowledge artefacts not as products of a mind situated elsewhere but as embedded within the vitality of matter, or within the creative practice of

Fig. 5.1 (a–c) Sensorial intra-actions with soil and embodied practices

preservation as an assemblage, turns the idea of soil, as well as food, paper or other materials, on its head.

The methodological concept of thing-power was unknown to us at the time, but in the workshop we deliberately set out to draw participants into an entanglement. Following Eisner (2003), we gave participants varying degrees of licence to draw and re-draw the lines between the factual and the fictional dimensions that foreground soil as a vital materiality. Looking back at the pictures, the vital force of soil is felt in the stretched-out hands of the participants. Soil left a pigmented trace on the skin, adhering to it and drawing out its moisture, with the hand opening out in response (Fig. 5.1a).

No longer positioned as inert objects of human volition and consumption, such materials are part of an agentic assemblage that includes, among its members, our own metabolism and cognition, as well as our moral sensibility (Bennett, 2010, p. 51). In this view, soil as thing-power cannot be contained within the abstract space of the head. It diffracts into a cartography of strips of paper, annotations, and scribbles, embodying an ongoing set of correspondences between bodies and materials, in the enactive and materialised expression of "a confederation of throbbing relations" (Bennett, 2010, p. 72).

One group found the relationships so entangled that their reassembling process resulted in the weaving together of the different elements (Fig. 5.1c), while others re-composed their collective experiences into a performance collage (Fig. 5.2).

The participants read the collage aloud in a rhythmic, rap-like fashion, combining body movements with varying tonalities and voices; re-mattering the force of soil as and through the embodied, transformative force of re-writing the soil into teachers' imaginations and practices. Diffractive processes led to ideas of making-with (Haraway, 2016) and to the possibilities of generating and prefiguring relationships that did not exist before.

The Anomalous Practice of the School Garden

Our second diffraction focuses on current experiences of working with children and young people in primary and secondary schools within a STEAM garden. This diffraction emerged through the relationships that arose from the departure from the controlled environment of the

Fig. 5.2 Workshop participants presenting their reassembled ideas about soil

classroom, traditionally associated with "learning", to the moment when students became involved in setting up their school garden.

In many respects, like other spaces within the school, the garden is traditionally associated with the teaching of specific curriculum subjects; it is used as a context to illustrate particular concepts in biology or the visual arts, with literature reporting increased levels of students' understanding as a result (Walshe et al., 2024). However, less common is the idea of approaching the garden as a form of "undercommon" space, which, following Harney and Moten (2013), may afford forms of everyday sociality. These spaces straddle the formal and informal, the vocational and the academic, yet they are also marginal within school education. For example, in schools, the garden may operate as an undercommon by extending beyond its functional uses to illustrate preselected concepts and to afford multiple relations, teasing the tension between work and leisure, cognition and

Fig. 5.3 (a and b) Primary school children in the school garden (reprinted from Gray et al., 2019)

emotions, and the common and diverse needs of humans and more-than-human others (Fig. 5.3a and b).

Hence, unlike those practices that are already saturated in the free play of social life, the school garden can be more effectively associated with what Harney and Moten (2013) call "practices of study", in the sense that they "actively resist the call to order imposed by forms of pedagogy, curriculum, policy theory or logistics" (Rousell, 2021, p. 66). Such practices unfold in spaces of non-compliance, in a dialectical, playful manner, rather than in non-compliance with the norms of the school, bridging intellectual and imaginative dimensions to give life, as Bennett (2010) argued, to the unexpressed and the denied possibilities, thereby letting education as a prefigurative practice begin.

As we look back at the images and the memories of the experience come flooding in, we recall the intensity of the garden's call, which grew week after week as the children's work in the garden took hold. Becoming familiar with the soil through sensorial exposure and encounters with the range of plants and animals (toads, snails, insects, etc.) generated moments of intense relationship with more-than-human others; it also became apparent that the children developed a different relationship with each other, entering new forms of non-linear co-causality and agency in the educational garden space:

> As the plants continued to grow, so did the children's status in the school, from pupils and learners to curators-guardians-gardeners-cultivators. In the

> acquisition of skilled practice there was a cultivation of an acquaintance, of a relationship of familiarity, of kinship. (Burnard et al., 2022, p. 188)

Skilled practice manifested itself as an ongoing attention to the garden as a space of encounters, as well as to its preservation as a practice of renewal and preservation (Verger et al., 2024), making the garden "felt and visible" in the co-adjacent space of the school. Children named their garden and developed a personal logo for it; they described the events in which they were both witnesses and participants, and they wrote about their garden in their school newsletter (Fig. 5.4a and b).

As Anna Gosetti-Ferencei (2018) discussed, in the life of the imagination, in art, learning is validated by the recipient's response and the collective response across a culture or community—in this case, the community of the school, the children themselves, and the plants. However, this also means that such validation carries power within the scientific community, if by science we do not simply refer to the consolidated body of professional knowledge but also to the capacity of humans to rethink themselves as both subjects and objects in an ongoing, reciprocal inquiry. As Bennett

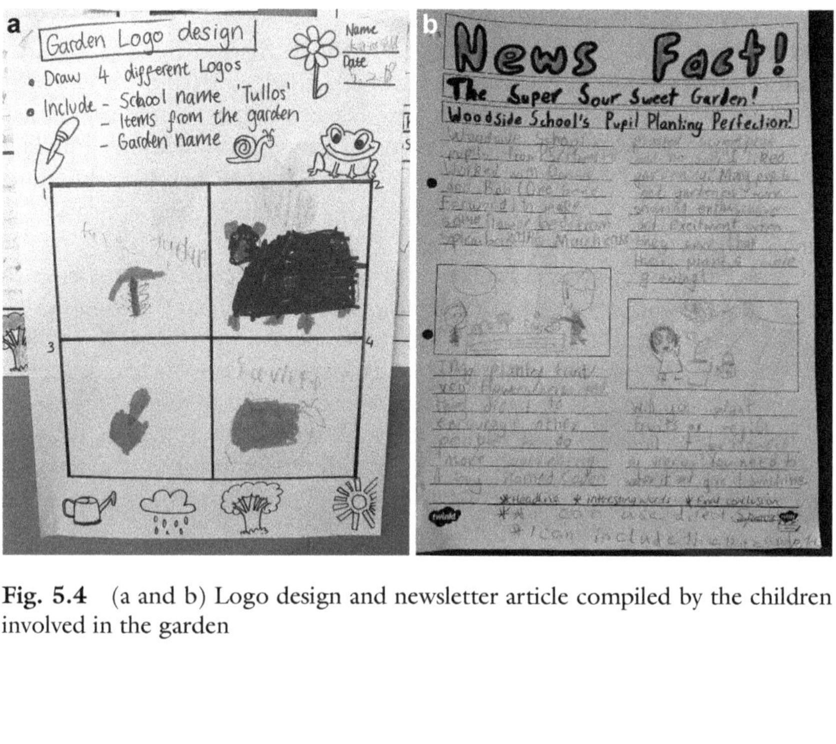

Fig. 5.4 (a and b) Logo design and newsletter article compiled by the children involved in the garden

(2010, p. 16) remarked, "humans encounter a world in which non-human materialities have power, a power that the 'bourgeois I', with its pretensions to autonomy, denies". We thus suggest that the relational ecology perspective of the STEAM garden is paramount in bringing young people into contact with their own and other people's lived experiences, and in challenging linear discourses of the production and consumption of both knowledge and resources.

Becoming with Soil: Possibilities in the Garden

The third diffraction brings us full circle, as the current project involves us as visiting researchers in a secondary school, working with a group of students and their home economics teacher. The first two diffractions brought to light the vibratory patterns of a configuration of throbbing relationships, as Bennett (2010) would have it, with the power to reorient the attention of teachers and students to each other and to the vitality of soil involved in embodied practices of making-with. In the context of the secondary school, the project evolved from an original partnership funding scheme available from the Royal Society for schools and a university partner to develop STEM educational programmes. Firmly situated within the vocational realm of formal education, the garden that sat—out of sight—at the back of the school was identified as a context for applications, data collection and monitoring of the ability of humans to make the unproductive space into a productive system.

In a second turn, with a different stream of funding awarded by Horizon Europe, the transdisciplinary figuration of the STEAM garden made other possibilities available. Part classroom, part kitchen, part laboratory, part hanging-out corner: both the home economics room and the garden showed the characteristics of Winnicott's (1989) transitional spaces, not simply in the physical sense of movement but of transit, that is, moving across different milieus, and capable of putting them into relation through playful inquiry and experimentation. Yet the conditions for transitional spaces are not just "there", as school spaces are normatively and socially codified, but they can be "staged" by operating, as Rousell (2021, p. 66) argued, through aesthetic acts "that environmentally redistribute the possibilities for bodily participation and affective engagement".

In this sense, the garden and the home economics classroom, conceived both as science and as art, became the primary context for a "practice of study" (Harney & Moten, 2013) situated between the model of the lab and that of the atelier (Hancock et al., in press). Here, we describe the careful, attentive, and persistent gesture of redirecting attention to the vitality of the garden by integrating scientific and artistic practices of attentive observation within dynamic entanglements (Fig. 5.5 a–c).

We invited students to go out regularly to the garden to collect leaves, flowers, and other specimens to decorate plates, make prints on tablecloths (Fig. 5.5a and b), or take inspiration for baking buns and biscuits in the shapes of vegetables and plants in the home economics base, the garden materialising itself and mattering in the hands of the students. Yet, within the home economics space, we also invited students to cook plant material to extract inks, which were then re-materialised as figurative plants (Fig. 5.5c).

As one of the girls observed: "it is so cool that the colour shows when it touches the paper ... you have to go very slowly ...". Making-with in this third diffraction was not simply an act of improvisation but a careful, attentive and persistent gesture of redirecting attention towards the quest for one's own and others' sustainability. In this view, critical thinking, normally associated with rational and cognitivist approaches, especially in science, is reframed as the enactivist process of sympoiesis or "making-with", whereby humans and non-humans participate in the same way in the process of knowing. By reconfiguring the human as part of the multiple relational assemblages of hands, soil, wind, leaves, seeds, seedlings, decaying litter, humus, light and water particles, engaging with the power of the "thing" does not imply taking a position from above but engaging in

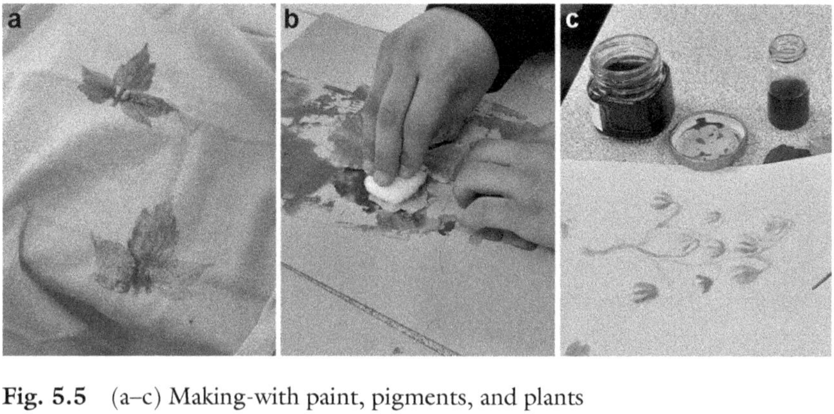

Fig. 5.5 (a–c) Making-with paint, pigments, and plants

practices through which "the world is differently articulated and accounted for" (Barad, 2007, p. 149).

So following Bennett (2010, p. 15), such sensory attentiveness may enable the exercise of an unrealistic imagination: one in which the high tables of the classroom become tables for a dinner party; the pots and pans work at once as expressions of the family home, speculation of a future profession and their ultimate reinvention as thing-power apparatuses. The vital materialities of the garden re-actualise themselves in the classroom space, as a variety of forms in aesthetic arrangements, in themselves clues to the vital correspondences they share (Fig. 5.6).

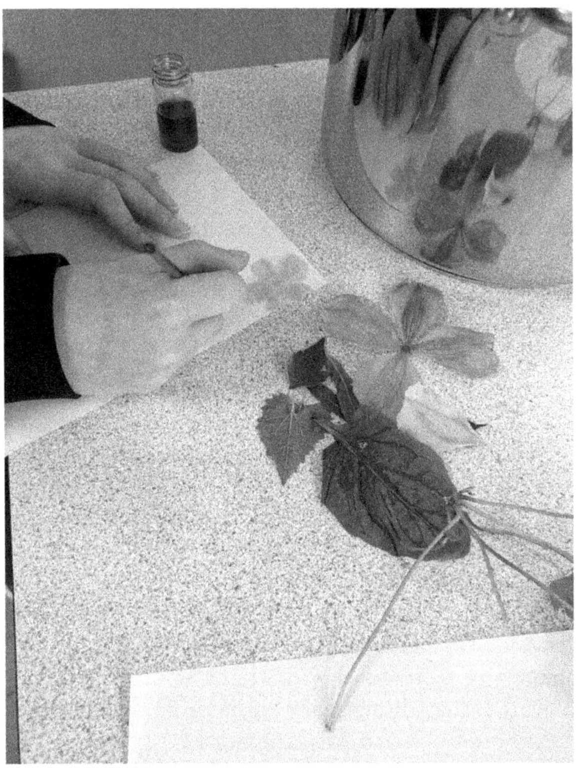

Fig. 5.6 Attentive observation: painting with freshly extracted botanical pigments

All these are mark-making gestures, lines of flight, bearing the possibility of possibilities. In this sense, the cartographies of the garden are cartographies of becoming and making-with the materials, places and bodies; they are also instances of reconfiguring educational practices as spaces of multiple possibilities. And so we ask: to what extent do we invite affective experiences in the science classroom, letting them guide the reshaping of subject boundaries and the remaking of new forms of co-living?

Evidencing the Materiality of "Thing-Power" in Mathematical Art Practices

Similarly, "thing-power" is evident in mathematical art practices where realities are mutually reliant. The interconnectedness of mark-making, inscribed gestures, and the visual aesthetics of mathematical art, together with the "mutuality" and "entanglement" of disciplines, makes reading them a disruptive act. Materials matter. Bodies matter. Multi-species coexistence matters. Seeing and re-seeing the performativity and emerging research apparatuses of mathematical art practice matters.

The two images that follow are from a set of 200 images by South African secondary school students featured in a publication *MathArt expressions by South African youth: Creative visual connections with mathematics* (GMMDC MathArt Team, 2022), published by a team of international researchers led by Dr. Carine Steyn and Professor Werner Olivier at Nelson Mandela University. This endeavour gave rise to a "MathArt methodology" based on our collective and emergent understandings that evolved during the analytical process. During these discussions, the term "MathArtWork" emerged, encompassing learner-produced, culturally situated, problem-inspired responses to how mathematics and art together contribute to knowledge. In order to offer some diffractive play with the "thingness" of "MathArt" materiality, we will feature an analysis of Fig. 5.7: Euclid's *The stressed Vitruvian man* and Fig. 5.8: Annika's *Soul number*. Both drawings trouble the ways the body is positioned and constituted in the material enactment of mathematics and art.

As we engaged in diffractive re-readings of these MathArtWorks, we were struck by the transdisciplinary interconnective meetings authored and emphasised here by South African youth.

Euclid's *The stressed Vitruvian man* (Fig. 5.7) shows the reality of maths. Euclid said:

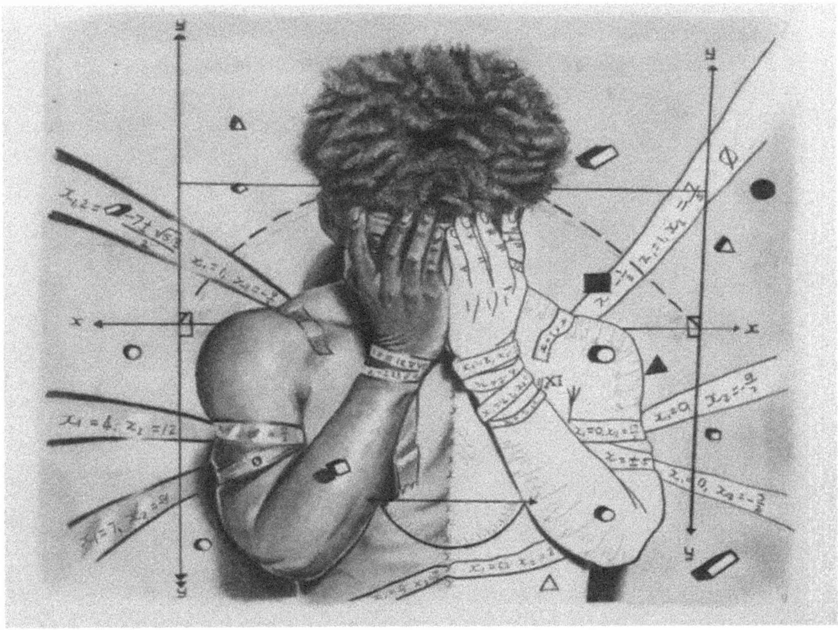

Fig. 5.7 Euclid's *The stressed Vitruvian man*

> We need measurements to make sure we get the right fittings ... I placed the equations on different parts of my body to show that there are different ways to get the answer. These things cause a negative toil inside my brain.

He focuses on himself, his hair, his hands, and his body, and shares how he thinks about the materiality of mathematics. The monotonality reflects different shades of black, with strong cultural references. The bi-tonal hands come to matter through cultural associations of anxiety, emotions, and bodily reactions, which take us into his world. We also see that the human body is the seat of mathematical knowledge and performs it. We connect with the things he has drawn and his experience of mathematics, which is inscribed on his body.

Annika's *Soul number* (Fig. 5.8) offers insights into her worldly spaces of school-home, everyday-extraordinary, material-conceptual, and production and performative understandings of practices of knowing. She communicates the complex and sophisticated mathematical concept of

Fig. 5.8 Annika's *Soul number*

infinity and number sequencing with dexterity, revealing a remarkable metaphorical quality and abstraction of things that portray her self-identities. We see the pre- and post-math-art self, manga self, South African self, student self, and math-artist self. The number lines in the form of flowing hair create visual texture, rhythm, and nuanced compositional definitions. Does this intimate that knowledge is "unending", hinted at also by hair drawn so that it seems to continue past the page? There is an abundance of symbolism and visual codes in the image. We begin by asking "why red"? Reflecting on the ubiquitous "thingness" of mathematics, Annika (in her self-narrative, which is too long to include here) acknowledges and considers how our lives cannot go one day without ... using numbers as things to represent or solve things.

So, what is the importance of "thingness" for methodology and interpretation in terms of the materiality of these drawings? Firstly, thingness catalyses different modes of thinking where bodies are in conversation with the material production of mathematical art. *The stressed Vitruvian man* (Fig. 5.7) invites us to re-see mathematical thinking through measurements inscribed on the body. Proportional relations and two shades of black make strong cultural references that come to matter through the bi-tonal hands.

Secondly, authors repeatedly but differently address "becoming" through mathematical art diffracted through the visual codes that feature in, for example, *Soul number* by Annika (Fig. 5.8), where we re-see number lines in the form of flowing hair that creates "unending" visual registers and the symbolism of infinity. Are her stitched lips a self-demonisation? Is the one red eye a symbolic self-reference to Annika's developing relationship with "being of the world" rather than "being in the world" (Barad, 2007, p. 160). Either way, hair has agency. Bodies have agency. Eyes have agency. Paper and pencils have agency.

Thirdly, "thingness" operates within entangled worlds of mathematics. In Figs. 5.7 and 5.8, we see the matter that matters, the precision that matters, in which all constitutive "things"—bodies, forms and faces, different shadings, algebraic expressions, analytic geometry, equations—work mutually and inseparably. We see the "thingness" of bodies and matter refocusing our attention, even a re-attunement with human–non-human worlds. We see the materiality of the human body, bodies of knowledge, and cultural meanings that are not passive but rather eruptively enacted through bodily reactions. "Thingness" is complex. "Thingness" counts as data.

These are methodological ways of seeing and re-seeing "thingness" that challenge the assumption that matter is passive. Rather, "thingness" is firmly rooted in the material world and works directly on the materiality of our bodies and the powers, forces, intensities and becomings which matter.

Evidencing the Materiality of Play as "Thing-Power" in Professional Western Art Music Performance

The "thingness" or "thing-power" of professional Western art music performance practices also offers a methodological window into the entanglement of music's materials in their intra-activity and effervescence. Whilst traditional music performance practices permit students to see over the threshold into and tentatively enter rigidly and externally defined spaces of control, an ontologically fluid "lens-shifting" approach to performance education engages with the multiple *possibilities* inherent in the materials of music learning and performance.

In 2025, Palmer et al. introduced the concept of "musicking-as-play", arguing for materiality as a primary domain of play in music. This is an ontological argument, using the speculative "as" lens to peer through the prismatic complexity of human musicking, turning it over in our hands to catch the movement of the light. What might the power of "as" be in sifting through the relational ontologies of music's materialities, its interplay of bodies (human and non-human), spaces, objects, and moving air? The materials of play are toys and, in his 1986 seminal exploration of toys as culture, Brian Sutton-Smith invites us to toy with the "as" lens: toys as culture, as bond, as obligation, solitariness, consolation, technology, education, market, agency, and so on. In doing so, he teases out the multiple and often conflicting uses of toys: as social gift and for solitary play; as treasured possession and as disposable trash; as offering autonomy whilst being "heavy with the missions of the culture of the toymakers" (1986, p. 215).

The ontological "as" recognises ontologies as fluid, contextual, relational, and useful. This is the same "as" that, from a theatre studies perspective, Gareth White employs when he writes about "game-as-art" (2024, p. 103). From an educational perspective, it entangles with ethics and epistemologies: how does this lens resonate with less-considered perspectives, and how might we know them through it? This strategy of insightfully "re-configuring" play's materialities extends to music through "musicking-as-play", an ontological lens that empowers those seeking to reconfigure music education for an uncertain future.

The concept of "play" resists definition, but a useful lens is to view it as a form of adaptive behavioural learning (Bergen, 2015; Brown, 2009, p. 49; De Koven, 2014, p. 151; Schore, 2008, p. 118), a model of agentic, intuitive, and personalised possibility studies that become hard-wired in the playing agent. Adaptive behavioural learning is a rhizomatic form of enquiry that constantly references the known-experienced against the known-thought and the unknown, building connections and alliances, filling in spaces and extending feelers. Its inefficiencies in relation to unilinear targets are outweighed by its self-making, future-crafting, habit-challenging, and "present-living" capacities. This is learning as play and play as learning: an ontological "flip" at the root of a possibility-charged education.

What material ecologies of possibility might there be in Western art music education, and how might we use them to open up practices of learning for music learners, for whom musical development is an experiential "coming-to-know" of the possibilities within the materialities of bodies, instruments, sounds, notations, relations and spaces? Can we enact "play" as an enquiry into finger dexterity, vibrating reeds, resonant spaces, wrist movements, dynamic shapes, rhythmic complexity, metal tubes and ensemble synchronicity? As Wilson suggests, music learning "itself becomes a practice of understanding, changing notions of matter" (2021, p. 1). Thomas Henricks (2015, p. 43), one of the foremost living play scholars, argues that such transformation is a defining feature of play, and so "musicking-as-play" is a "making things happen", an empowering materialisation of "thing-power" whether in our neurons, our little fingers, our violin bow, the vibrating air, or the concert hall. For Henricks (2015, p. 43), a second defining feature of play is its consummatory nature, meaning that, in bounded moments, the player is fully absorbed in sound-making through the material intra-play of human and non-human materials.

So, what then is the consummatory thing-power and "thingness" of "instruments-as-toys"? Toys absorb children *and* adults alike, designed to hook and hold the attention of the player as "instruments for letting play loose in the world" (Sicart, 2014, p. 47). Again, however, their paradoxes are manifest: toys are materials "designed and fashioned to be tangible, touched, and manipulated. Yet, they are also the least material of objects, tending in play to conjure and scaffold imaginary events and worlds and come to life in dreams and secret adventure" (Giddings, 2024, p. 11).

So toys model miniature worlds whilst sparking dreams of vast ones: there is a "persistent interrelationship of the material and the imaginary in play with objects" (Giddings, 2024, p. 52). However, the *materials of play,* such as the doll or the Lego brick, do not just express and represent cultural norms (Wilson, 2021, p. 53) but also provide the player with the tools to adaptively parody, critique, and symbolically manipulate that world (Giddings, 2024, p. 11; Sutton-Smith, 1986, p. 247). As with toys, so with instruments: pianism *as* routine practice or *as* sublime communication; the violin *as* the striving solo voice or *as* the relational tool of a uniform community of string players; the trumpet fanfare *as* the emblem of communal empathy or *as* the humorous critique of pomposity. Again, in our conceptualising, we are falling back on the "as" formulation: "x-as-y", ontological fluidity for the sake of insight into multifaceted phenomena.

Burnard and Köbli (2024, p. 252) remind us that "thinking of ontology as processual has significant implications for how we understand what research is". What then does looking at "instruments-as-toys" show us about research into thing-power in music? It demonstrates that the human focus of our educative acts extends agency into "thinking with/in" materials through partnerships with objects, spaces, resonances, and other resources (Bates, 2012, p. 364) that exemplify how "objects constitute extensions of the self" (Diesendruck & Perez, 2015, p. 18).

This forms an intra-active assemblage that is emergent in each temporal instance, in which "instruments-as-toys" make their makers and users as much as they are made by them, and in which the body-in-performance—neurons, adrenaline, lungs, embouchure, digits—is a fundamental constituent part. The inscribed material affordances of an instrument mean that they "may know things for their users" (De Souza, 2017, p. 24) in how they "discipline the body into culturally approved configurations" (Cook, 2018, p. 50). A handcrafted professional musical instrument exhibits a material variability which contributes an active voice to the assemblage, in the grain of its wood or the hammering of its bell. Such assemblages craft identities and transformative directions for facing the world (Sutton-Smith, 1986, p. 207): "instruments-as-toys" are "transformed and invested with new meanings that reflect and assert who we are" (Attfield, 2020, abstract) and are imbued with a shared vitality which suggests "a lingering discomfort and inadequacy with human-centred conceptualizations of 'performance' and 'agency'" (Bates, 2012, p. 387).

In summary, musicking-as-play, with "play" here described as a form of adaptive behavioural learning with both transformative and consummatory characteristics, functions as an exemplar of the "ontological 'as'". This is a processual "lens" for thinking *through* varied ontological stances, employed here to (re)conceptualise how Bennett's (2010) "thing-power" might manifest in Western art music. Engaging the concept of "instruments-as-toys"—or the "thingness" of instruments—enables us to think through and with music's materialities as they draw together the human and non-human into emergent performative assemblages. These assemblages orient us differently towards questions of agency and identity, with the "instruments-as-toys" ontological lens revealing complex layers of meaning-making in relation to the human researcher and musical materiality, demanding revitalised research methodologies that decentre the human. Perhaps Bogost's suggestion that "play is a material property of all objects" (2016, p. 119) will help us reconsider the agentic possibilities of materiality, music's "thing-power" "*as play*" within music education research. Our relationship with soil.

Conclusion

In this chapter, we have argued that, methodologically, assemblages of thing-power fundamentally challenge the focus on linear trajectories defined by targets and end goals in educational practice. Instead, in line with Verger et al. (2024), we argue that other narratives, founded on materiality, play, and making-with thing-power, are possible by exploring and recombining the affordances inherent in what already exists. Through our three diverse experiences, we have illustrated that such narratives are not simply ideological or abstract but materially tangible. They stem from the complex assemblage of practices that humans, non-humans, and more-than-human co-author, and from the distributive agency of "thingness" that embodies all research practices, as evidenced here in STEAM gardening, mathematical art, and music performance practices. Taken altogether, our experiences account for the plurality of existence and *reconfigurings of educational practices* as spaces of multiple possibilities.

As Barad (2007, p. 160) writes:

> [W]hat is at issue and at stake is a matter of the nature of reality, not merely a matter of human experience or human understanding of the world. Beyond the issue of how the body is positioned and situated in the world is the matter of how bodies are constituted along with the world, or rather as 'part' of the world (i.e. 'being-of-the-world', not 'being-in-the-world').

So, what material ecologies of possibility might there be here in research methodologies more generally? If materials are *in play* and humans and non-humans alike are permanently bound up in a play-fuelled material emergence, how might we consider the play of research on such agents? Does the lens of "*research-as-play*" reveal potential for action by those who seek to explore conviviality, belonging, and extended possibilities for human and more-than-human relations in educational spaces and in educational research? Going beyond the economic frameworks of production, the ontological "as" offers a catalytic impetus for play-*full* investigations into the materialisations of "thingness" and thing-power across research domains; and in line with Verger et al. (2025), by valuing the existence and dignity of plurality as "multiple-things-in-the-world", we have called for research practices which are invitational of "as-yet-unthought" potentials for new knowledge but also new forms of habitation and co-habitation with diverse others in the world.

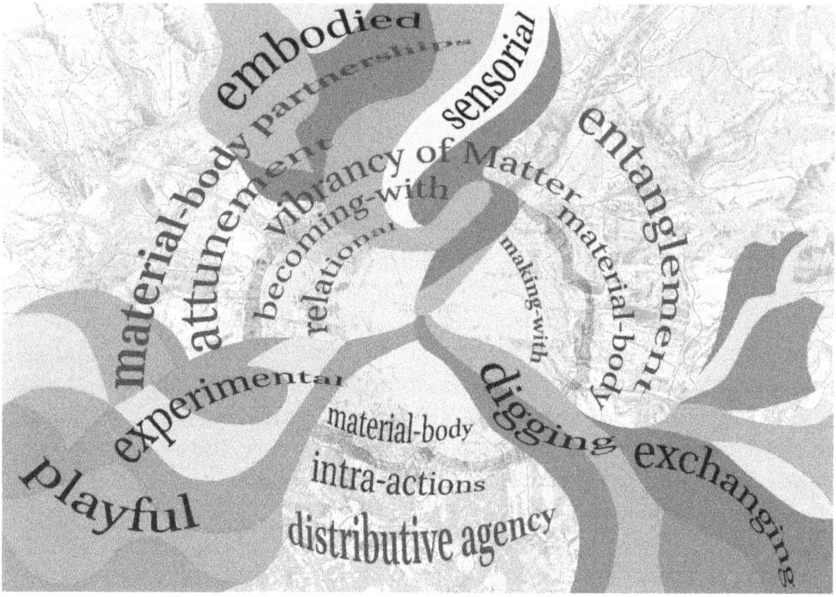

Fig. 5.9 An assemblage of concepts for mattering and materialising "thingness"

In light of this, instead of researching the form or representation of "things", we invite you to consider the assemblage of concepts in Fig. 5.9 for researching how "thingness" maps out across intra-acting objects and agencies. How might this cartographic assemblage of theories and concepts give rise to new timely discourses about the co-existence and mutual co-production of alternative educational possibilities?

What might re-mattering "thing-power" and "thingness" generate as alternative ways of producing knowledge for enacting societal change? We invite you to reset the significance of "things" that seem to have only marginal importance and allow them to matter as shared practices of collective, everyday experiences.

References

Attfield, J. (2020). *Wild things: The material culture of everyday life* (2nd ed.). Bloomsbury.

Barad, K. (2007). *Meeting the universe halfway: Quantum physics and the entanglement of matter and meaning*. Duke University Press.

Bates, E. (2012). The social life of musical instruments. *Ethnomusicology, 56*(3), 363–395. https://doi.org/10.5406/ethnomusicology.56.3.0363

Bennett, J. (2010). *Vibrant matter: A political ecology of things*. Duke University Press.

Bergen, D. (2015). Psychological approaches to the study of play. *American Journal of Play, 7*(3), 101–128.

Blenkinsop, S., & Kuchta, E. (2024). *Ecologizing education: Nature-centered teaching for cultural change*. Cornell University Press.

Bogost, I. (2016). *Play anything: The pleasure of limits, the uses of boredom & the secret of games*. Basic Books.

Braidotti, R. (2019). *Posthuman knowledge*. Polity.

Brown, S. L. (2009). *Play: How it shapes the brain, opens the imagination, and invigorates the soul*. Avery.

Burnard, P., Colucci-Gray, L., & Cooke, C. F. (2022). Transdisciplinarity: Re-visioning how science and arts together can enact democratizing creative educational experiences. *Review of Research in Education, 26*(1), 166–197. https://doi.org/10.3102/0091732X221084323

Burnard, P., Colucci-Gray, L., & Sinha, P. (2021). Transdisciplinarity: Letting arts and science teach together. *Curriculum Perspectives, 41*(1), 113–118. https://doi.org/10.1007/s41297-020-00128-y

Burnard, P., & Köbli, N. A. (2024). Posthumanist new materialist pathways for reimagining music education research: What matters? What can this offer music educators? *Music Education Research, 26*(3), 251–263. https://doi.org/10.1080/14613808.2024.2350470

Colucci-Gray, L. (2022). The STEM, STEAM, STEAME debate: What does each term mean and what theoretical frameworks underpin their development? In J. Dillon & M. Watts (Eds.), *Debates in science education* (pp. 13–26). Routledge.

Colucci-Gray, L. (2025). Carbon entanglements: For a science education in the space of the sensible. *Policy Futures in Education, 23*(3), 553–567. https://doi.org/10.1177/14782103251320828

Cook, N. (2018). *Music as creative practice*. Oxford University Press.

De Koven, B. (2014). *A playful path*. Carnegie Mellon University. https://doi.org/10.1184/r1/6686705.v1

De Souza, J. (2017). *Music at hand: Instruments, bodies, and cognition*. Oxford University Press.

Deleuze, G., & Guattari, F. (2013). *A thousand plateaus*. Bloomsbury.

Diesendruck, G., & Perez, R. (2015). Toys are me: Children's extension of self to objects. *Cognition, 134*, 11–20. https://doi.org/10.1016/j.cognition.2014.09.010

Eisner, E. W. (2003). On the differences between scientific and artistic approaches to qualitative research. *Visual Arts Research, 29*(57), 5–11.

Fox, N., & Alldred, P. (2017). *Sociology and the new materialism: Theory, research, action*. Sage.

Frank, A., Gleiser, M., & Thompson, E. (2024). *The blind spot: Why science cannot ignore human experience*. MIT Press.

Funtowicz, S., & Ravetz, J. R. (1994). Uncertainty, complexity and post-normal science. *Environmental Toxicology and Chemistry, 13*(12), 1881–1885. https://doi.org/10.1002/etc.5620131203

Giddings, S. (2024). *Toy theory: Technology and imagination in play*. MIT Press.

Giuffredi, R., Colella, C., Colucci-Gray, L., Caretto, A., Spagna, R., & L'Astorina, A. (2025). Beyond crisis talk: Making time for researching new narratives of human relations with soil. *Tecnoscienza*, 33–53. CNR Institutional Research Information System. 10.6-92/issn.2038-3460/18076.

Glăveanu, V. P. (2023). Possibility studies: A manifesto. *Possibility Studies & Society, 1*(1–2), 3–8. https://doi.org/10.1177/27538699221127580

GMMDC (Goven Mbeki Mathematics Development Centre) MathArt Team. (2022). *MathArt expressions by south African youth: Creative visual connections with mathematics*. Nelson Mandela University.

Gosetti-Ferencei, J. A. (2018). *The life of imagination: Revealing and making the world*. Columbia University Press.

Gray, D., Colucci-Gray, L., Donald, R., Kyriacou, A., & Wodah, D. (2019). From oil to soil. Learning for sustainability and transitions within the school garden: A project of cultural and social re-learning. *Scottish Educational Review, 51*(1), 57–70. https://doi.org/10.1163/27730840-05101007

Hancock, J., Colucci-Gray, L., Darling-Macquistan, K., & Chapman-Kelly, S. (n.d.) (Accepted/in Press). Decolonising primary teacher education through embodied and sensory approaches in STEAM education. *Scottish Educational Review*. Special Issue: Decolonisation in Teacher Education.

Hancock, J., Gray, D., & Colucci-Gray, L. (2025). A view from the garden: Interrupting politics of attainment and re-orienting education towards sustainable futures. *European Educational Research Journal, 24*(1), 68–86. https://doi.org/10.1177/1474904123118996

Haraway, D. (2016). *Staying with the trouble: Making kin in the Chthulucene*. Duke University Press.

Harney, S., & Moten, F. (2013). *The undercommons: Fugitive planning & black study*. Autonomedia.

Henricks, T. S. (2015). *Play and the human condition*. University of Illinois Press.

Jackson, Z. I. (2020). *Becoming human: Matter and meaning in an antiblack world*. NYU Press.
Lury, C. (2020). *Problem spaces: How and why methodology matters*. John Wiley and Sons.
Meskens, G. (2024). The ethical motivation for post-normal science. *Futures, 163*, 103457.
O'Sullivan, S. (2024). *From magic and myth-work to care and repair*. Goldsmith Press.
Palmer, T., Burnard, P., & Burke, D. (2025). Inviting a (re)orientation to 'musicking-as-play' in higher music education performance studies: Insights from three genre performance practices. *Action, Criticism, and Theory for Music Education, 23*(2), 7–47. https://doi.org/10.22176/act24.1.7
Rousell, D. (2021). *Immersive cartography and post-qualitative inquiry: A speculative adventure in research-creation*. Routledge.
Saltelli, A. (2024). Scienza post-normale. In *Enciclopedia Italiana di scienze, lettere ed arti* (Appendix 11). Istituto della Enciclopedia Italiana.
Schore, A. N. (2008). Playing on the right side of the brain: An interview with Allan Schore. *American Journal of Play, 9*(2), 105–142.
Scoones, I. (2015). Transforming soils: Transdisciplinary perspectives and pathways to sustainability. *Current Opinion in Environmental Sustainability, 15*, 20–24. https://doi.org/10.1016/j.cosust.2015.07.007
Sicart, M. (2014). *Play matters*. MIT Press.
Sutton-Smith, B. (1986). *Toys as culture*. Gardner Press.
Taguchi, L. H., & Palmer, A. (2013). A diffractive analysis of the performative enactments of girls' ill–/well-being with(in) school environments. *Gender and Education, 25*(6), 671–687. https://doi.org/10.1080/09540253.2013.829909
Verger, N. B., Duymedjian, R., & Glăveanu, V. P. (2025). The meritocracy of preservation: Reimagining merit beyond production. *Organization*. Advance online publication. https://doi.org/10.1177/13505084251347519
Verger, N. B., Duymedjian, R., Wegener, C., & Glăveanu, V. P. (2024). Creative preservation: A framework of creativity in support of degrowth. *Review of General Psychology, 28*(3), 268–281. https://doi.org/10.1177/10892680241256312
Walshe, R., Evans, N. S., & Law, L. (2024). School gardens and student engagement: A systematic review exploring benefits, barriers and strategies. *Issues in Educational Research, 34*(2), 782–801.
White, G. (2024). *Meaning in the midst of performance: Contradictions of participation*. Routledge.
Wilson, S. J. (2021). *New music and the crises of materiality: Sounding bodies and objects in late modernity*. Routledge.
Winnicott, D. W. (1989). *Playing and reality*. Routledge.

CHAPTER 6

Rules for Researching: Rules for Possibility

Harriet Hand

Introduction

On my wall there is a postcard. It is red. Across the centre are three words: "everything is interesting."[1] It is a reminder, an encouragement. It is a rule.

This chapter examines methodologies for possibility thinking. Perhaps it seems incongruous; I argue that rules are a productive technique for fostering possibility thinking. In fact, rules are essential if we are to move beyond practised ways of thinking and doing. I make my argument by exploring how a set of custom rules that guided me through my postgraduate research study created the conditions for possibility.

Rules are everywhere (Daston, 2022). As I navigated postgraduate study, I encountered institutional, school, grant, ethical, and methodological rules and, due to the ongoing COVID-19 pandemic, a whole lot of new rules that caused havoc with all of these others. The rules I discuss in

[1] This postcard was sold as part of an exhibition by the artist Kelly Mark at the Ikon Gallery in Birmingham in 2003.

H. Hand (✉)
School of Education, University of Bristol, Bristol, UK
e-mail: harriet.hand@bristol.ac.uk

N. B. Verger, D. Mattson (eds.), *Possibility Studies and Societal Futures*, Palgrave Advances in Possibility Studies,
https://doi.org/10.1007/978-3-032-26379-7_6

this chapter are a set of custom rules that I developed in the early stages of my PhD. At first glance, these rules read as simple instructions. As the chapter will show, exploring how these rules performed is a way of exploring a more complex story about the nature of rules and the work that rules do. I hope to show how rules were integral to how possibility was made in the research event. In a necessary and continual endeavour to broaden and deepen our understanding of the complexity of possibility thinking (Glăveanu, 2023), I am curious as to how rules might just be what we need—not to keep us on the straight and narrow but to encourage us off the beaten path.

My PhD inquiry explores the intersection of creative thinking, process philosophy, and the emergent and dynamic properties of participatory mapmaking as a world-making practice (Gerlach, 2014). In the context of the inquiry, mapmaking refers to the everyday practice of bringing different visual elements—text, image, or abstract expression—into relation on paper. The emphasis is on the process and practice of making maps, rather than the final map as an object of representation. The inquiry engages with the notion of mapmaking as an emergent practice; therefore, my interest is in how it brings forth new ideas about self and the world. I engage an image of thought proposed by Deleuze (1994/1968) and elaborated in collaboration with Guattari (Deleuze & Guattari, 2013/1988). This image suggests *genuine thought* as an everyday complex phenomenon that creates new versions of self and world through a relational process of cognitive, embodied, and sensed encounters (Deleuze, 1998). Grounded in this process philosophy, the project explores the question of what it means to create conditions for genuine thought—or rather the as-yet-unthought—to flourish. It is concerned with the conditions that permit and activate possibility thinking, and I use the practice of participatory mapmaking as a way of exploring these conditions.

The project was centred around a mapmaking residency in a space within the Watershed—a cultural centre in Bristol, UK—during six weeks in the Spring of 2022. Weekly sessions with nine 17-year-old students resulted in maps and other artefacts that came to cover five large display walls that contained the space. A continual flow of visitors provided critical engagement with what emerged; comments were added on sticky notes, new questions were asked, and the artefacts we made moved around the room.

Over the course of the research, I came to see mapmaking as a practice that dissolves the framings that commonly separate self and the world to give access to, and enliven, the complexity of daily life. In my thesis (Hand, 2024), I explore how mapmaking practices lend themselves to the disruption of framings that stifle experience and limit expression because I observe how maps emerge as a result of *feeling* and *sensing* relationships rather than *planning* connections. Thought is furthered through disruptions that challenge assumptions of place, experience, or memory that have been rendered static over time. I introduced the idea of an unreliable pedagogy in a public talk during the residency as a practice of teaching and learning that gives no assurance as to what will be produced (Watershed, 2023). In this sense, mapmaking becomes a pedagogic practice that is transformative because it is deeply situated (Lave & Wenger, 1991) and embedded in experience (Dewey, 1963), accounting for the senses, emotions, and sense of belonging. The conditions mapmaking creates constitute a learning space of anticipation and curiosity that holds promise for sustaining the possibility of change.

In our mapmaking space there was no assurance that anything was fixed, no guarantee that activity would stay neatly within any boundaries that were set, and no blueprint for us to follow. Attempting to construct such a space—and I do not claim we always got this right—took a fair amount of grappling with the question of what it means to research when the problem you are seeking to explore is one of permitting, furthering, or activating the complex dynamics of everyday life (Vannini, 2015).

Embarking on this research journey, I took an interest in methodologies that explore the concepts of Deleuze and Guattari (see, for example, Alvermann, 2000; Leander & Rowe, 2006; Vagg, 2022). I found myself aligned with a post-qualitative approach to research that challenges methodologies that are founded on a principle of reporting, describing, or validating any kind of "truth" (St. Pierre, 2021). Non-representational theory and methodologies resonated with me for the way they share an interest in relationality, complexity, and emergence (Boyd & Edwardes, 2019; Thrift, 2008; Vannini, 2015). I was encouraged by speculative approaches to researching (Savransky & Stengers, 2018; Savransky et al., 2017), and I became drawn, in particular, to research-creation (Truman et al., 2020) for the way it embeds creative and experimental practices in its application (see also Burnard et al., Chap. 5, this volume).

There are important distinctions amongst these approaches, but what felt important is their interest in "what is made possible" (Mazzei & Jackson, 2023, p. 170). Aligned with Deleuze's image of thought (1994/1968), this interest encourages an openness to what might lie beyond the boundaries of the problem at hand, a challenging of what makes sense to us, and a risk-taking with the thoughts and feelings that bubble up in the moment before we can fully grasp them.

To orient myself to research in this way, I explored ways of allowing the study to unfold as a process rather than be planned out as a procedure. I was drawn to methods that account for, and productively engage with, the complexity that this gives rise to. For example, I was drawn to the use of productive provocations by Stephanie Springgay and Sarah Truman (2017). These short texts are reminiscent of Fluxus artist instructions that are used as a method for making artistic performances that blur the boundaries between art and life (Higgins, 2001). In my study, I used similar provocations that worked to activate participation in the research event without dictating how this might look and feel.

MAPPING #034 Everyday

Draw a map of your everyday. Consider the places. Make paths. What are the landmarks? Find the edges.

I was also drawn to Maggie MacLure's notion of wonder (2013b; see also Mattson, Chap. 11, this volume). This idea encourages a sensibility to things that call your attention even when you can't explain it—a form of knowing that is lost when we limit ourselves to conventional systems of engagement based on language alone. Techniques and exercises such as these encourage an engagement with what is being produced and sensed in the moment of encounter. They are generative without asking us to think too much about whether what is being produced makes sense, or where it might lead.

The problem of rules came up as I was grappling with questions of method. Rules and research are inextricably linked. There is an abundance of rules that can be found in relation to all stages of methodology: in selecting and recruiting participants; in carrying out ethical research; in analysing data; or in following conventions for reporting. As a new postgraduate researcher, the idea of "rules" felt instinctively attractive. Feeling more secure about my orientation to the research, the idea of having some rules felt useful as I moved from theory to practice. However, the dominant rules that I saw around me were not suited to how I was orienting to the research. It is the open-endedness of social science research that asks us to apply rules that help surface the unexpected, the contrary, the different (Firebaugh, 2008). But these kinds of rules do not come ready-made. To see the world differently, rules must take account of the situation at hand and be born out of the conditions of the research event itself.

Rules for Researching

Early in my studies, a particular passage in *A thousand plateaus* (Deleuze & Guattari, 2013/1988, p. 26) prompted me to think about the idea of rules and the kind of rules that might be productive in my research practice.

> Write with slogans... Make rhizomes, not roots, never plant! Don't sow, grow offshoots! Don't be one or multiple, be multiplicities! Run line, never plot a point! Speed turns the point into a line! Be quick, even when standing still!

I was struck by the tone of these lines and how it produces a particular rhythm: urgent, punctuated, a rallying cry: *Write! Make! Grow! Be quick!* These were instructions I needed to take seriously. These were rules.

These kinds of rules were not new to me. What came to mind most readily were the *Ten rules* of Sister Corita Kent (1918–1986), an artist, educationalist, and activist for social justice (Makelburg, 1967). For Kent, art was not about expression; she saw art as an integral part of being human and experiencing the everyday. Kent's *Ten rules* are pedagogic rules that encourage students and teachers to remain present to what is happening, resisting the urge to fix on a particular meaning of the past, in order to create new worlds (Lanier, 1965).

What the *Ten rules* do for me as a teacher is help me see the reciprocity between teacher and student. Her rules say: *Students! Pull everything out of your teacher. Teachers! Pull everything out of your students.* I shared this idea with a group of students; we reflected on the responsibility to be active and curious as learners — a sense that "we are in this together." Kent's encouragement to be present to the instability and possibility of life — and to approach it as an experiment — resonates with my own experience of art education. The rules are full of energy; they are specific, yet they feel liberating. In this sense, they act as a model for creating rules for researching because they have all the generality that helps them be applied in a multitude of situations, and all of the embodiment of a particular orientation to life and learning.

Another text that caught my attention, as I was finding my way into the methodology of my study, was the manifesto *First things first*, instigated by the designer Ken Garland in 1964. In the spring of 2021, at the start of my studies, I learned that Ken Garland had died (Shaughnessy, 2021). I recalled reading his manifesto for the first time during my undergraduate studies in design in the late eighties. Now, I found myself returning to it and reflecting on its force as an incitement to the design community not to be implicated in the damaging work of commercial enterprise and instead direct efforts and skills towards more worthy environmental, social and cultural causes. *First things first*, as a manifesto, does not take the form of a set of instructions, but its dual function, both as a call to action and an expression of a particular worldview, means it is operating along very similar lines to Sister Kent's *Ten rules.*

This feels like a good moment to turn to my own rules for researching. The following is the list of rules and a photograph of the section of wall where they hung in the research space (Fig. 6.1).

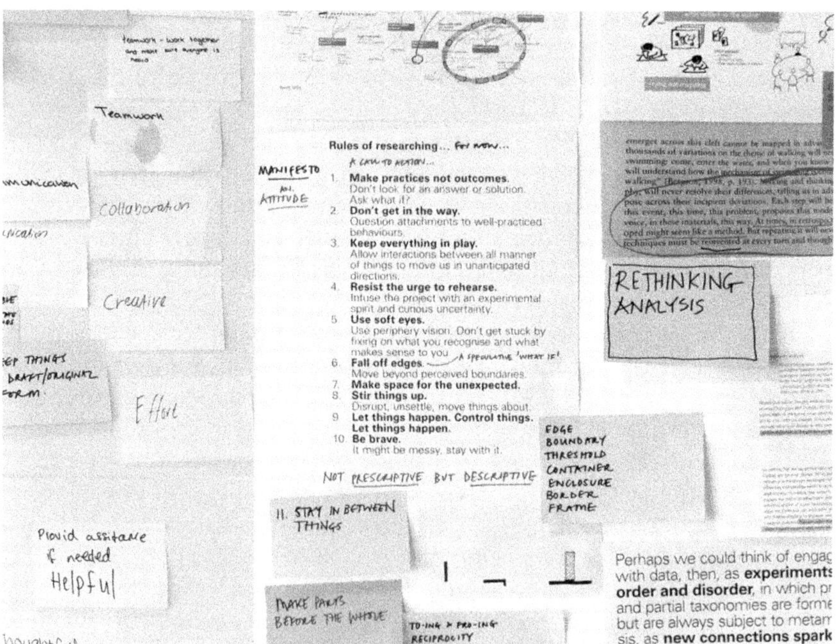

Fig. 6.1 The rules for researching displayed on the walls of the research space

Rule 1: Make practices not outcomes.
Rule 2: Don't get in the way.
Rule 3: Keep everything in play.
Rule 4: Resist the urge to rehearse.
Rule 5: Use soft eyes.
Rule 6: Fall off edges.
Rule 7: Make space for the unexpected.
Rule 8: Stir things up.
Rule 9: Let things happen. Control things. Let things happen.
Rule 10: Be brave.

Like Corita Kent's *Ten rules* or Ken Garland's manifesto, these rules for researching speak to a particular orientation; they are grounded in the theory I was reading and a reflection of what felt important to me as my research study began to take shape. Sometimes, the wording for a rule came directly from the texts that were guiding me; in other cases, the rule emerged through a more complex and messy coming together of ideas, feelings, hopes, fears, or memories. Daston describes rules as a "betwixt and between" category telling us that rules sit somewhere between "guidelines that could be taught and savvy that could only be acquired through practice" (2022, p. 4). Such a characteristic seems well-suited to post-qualitative research and the desire to blend the distinctions between theory and practice (Jackson & Mazzei, 2018). I will give a brief description of how this blending works in the context of the rules for researching and introduce how each rule came into being.

I start with **Rule #3 Keep everything in play**, because it has echoes of the postcard I describe at the start of this chapter. The rule came about through thinking with the concept of multiplicity (Deleuze & Guattari, 1994/1991). Following Spinoza, Deleuze and Guattari use multiplicity to describe how the world comes into being through an unfolding process of connection and disconnection between all manner of things that are not constrained by any single external or determining force. They suggest that: "To attain the multiple, one must have a method that effectively constructs it" (Deleuze & Guattari, 2013/1988, p. 23). Therefore, **Rule #3 Keep everything in play** felt important to the making of an open research space, where relational forces would have the greatest possibility of *constructing the multiple*. The rule also draws upon my arts-practice savvy that tells me all manner of things have the potential to further the process of creation. As in art, **Rule #3** permits participation in a particular kind of educational encounter: one where the sheer number and diversity of things that we consider matter are kept intentionally open as a process of exploration and possibility (Facer, 2016).

This same arts practice savvy tells me that participation in such encounters is not straightforward. **Rule #2 Don't get in the way** was conceived as an acknowledgement of my own tendency for organising and tidying up—which can mean that things can be left out in the process of exploring a problem. Sennett (2008) suggests that instructions can arise from anticipating difficulties that may be encountered; this rule keeps me mindful of the danger of familiar ways of working. Deleuze (1994/1968) helps me see that holding some sense of control over the conditions of thinking—by applying practised ways of working, for example—risks casting shadows over something that might not agree with previously held beliefs. These conditions of confusion and discomfort—that will endure if we don't get in the way of them—are exactly what Deleuze is encouraging as productive of genuine thought.

One way of enacting **Rule #2** was to use the technique of **Rule #5: Use soft eyes**.[2] The idea of *soft eyes* came up during a call with my supervisor, Jennifer Rowsell, and our colleague and friend, Mark Shillitoe. We were talking about our experiences of teaching during lockdown. Mark said: "It's like that line in *The Wire*." I was not familiar with the TV show, so Mark told me how the phrase relates to a kind of surveillance. I thought about the test I do in the opticians, where I push a button when I can see something flashing in my peripheral vision, and the way I soften my gaze to do this. The technique struck me as useful, and it gained weight as I began to think with Deleuze's concept of recognition (Deleuze, 1994/1968). Recognition can render a thought as "good" because it fits within our usual frames of reference for how things are. Observing **Rule #5: Use soft eyes** offered a way of challenging my natural instinct to be drawn to what makes sense to me; it works as a metaphor for an attention to the informal, the uttered, something that might not make sense, yet is expressing some kind of change (Lorimer, 2015).

As I was devising the rules, I also acknowledged a habit I have of playing things through in my head ahead of time, a habit of rehearsing. **Rule #4, "Resist the urge to rehearse"**, began to feel important to challenging such a habit; after all, how could I do speculative work if I had already assumed how the action would play out ahead of time? Or was it possible

[2] *Soft eyes* is the title of an episode of the US television series *The Wire* on NowTV.

to rehearse openness? Thinking with these questions, I came to see the habit of rehearsing as having less to do with ensuring a particular outcome and more about being vigilant to threats or risks — to either the people involved or to the process. *Rehearsing* became the idea of anticipating how I needed to care for both. **Rule #4** gets at something of an inner tension between my teacher-self being concerned about the confusion of an overly open environment (Cremin & Chappell, 2019) and my researcher-self, who is motivated by the way a speculative approach permits growth and change.

Rule #1: Make practices not outcomes came into play here. This rule felt important in encouraging an experimental attitude rather than anticipating any particular result. This rule is about exploring ways of making practices from within the process of the event and of reinventing these techniques as the action unfolds (Manning, 2015). The rule held me accountable to the kind of iteration familiar to my design practice, where solutions respond to the concrete situation and the needs of people encountering that situation.

The practical knowledge that encouraged **Rule #1** is the same knowledge that tells me practices are not about working with an entirely open set of parameters; **Rule #9: Let things happen. Control things. "Let things happen"** introduces the dynamic rhythm of paying attention to how things can be both open and constrained in different ways at different times. The form of the rule echoes the principle of *control and surrender*, a concept used by artist and musician Brian Eno in his creative practice (Marshall & Loydell, 2015). This practice assumes constraints as enabling rather than limiting, as parameters imposed on an encounter that do not pre-empt a particular outcome but offer opportunity for exploration (for different takes on constraints see, for example, Miles & Springgay, 2020 and Tromp, 2023).

By now, the gist of these rules is becoming apparent as a set of instructions for keeping the process of creation, of making possibilities, going. **Rule #7, Make space for the unexpected**, continues in this vein by encouraging opportunities for things to happen that we might not have seen coming. The rule has echoes of Anna Craft's notion of "standing back" (2015, p. 162), a move that is familiar to my teaching practice, observing the benefits of allowing students to find their own way through a problem. Later in the study, the rule gained importance as a guideline for me to consider the agency of the maps, the possibility that if we give the maps space and time—by keeping them on the walls and part of the action—possibility may emerge from the object itself (Rossetto, 2020).

Rule #7 might be guidance to step back, pause, or slow down, **Rule #8: Stir things up** is about giving permission for play, to move, to disrupt, to meddle. Similarly, **Rule #6**: **Fall off edges** is also concerned with encouraging a particular kind of attitude that might help jog us out of familiar frames. The idea of falling off edges came from reading a passage in the critical geographer Doreen Massey's book *For space* (2005). Massey suggests that falling off an edge—going beyond the limits of our comfortable or familiar spaces—is how we come upon difference; it encouraged me to think of edges as hopeful and lively spaces. Both **Rule #8** and **#9** felt important to keeping me from the danger of retracting to a safe space of known practices, where nothing other than repetition might be produced (Deleuze, 1994/1968). To close this brief overview of how the rules came to be, **Rule #10**: **Be brave** was a response to the other nine rules and a recognition of what it might take for me to observe them.

In the research space, the rules were displayed on one of the walls. I drew attention to them as a way of expressing the orientation of the study and how I was being guided—and how I wanted to encourage others to be guided. The rules came into conversation with the students and acted as an encouragement. The students also surfaced new guidelines that felt important—*Stay in between things*, for example, became **Rule #11** when the students noticed that comparing or contrasting two different artefacts was when something interesting or important seemed to happen (in the middle). However, the rules were not imposed on others because I am sensitive to how the rules are situated in the texts I am reading and thinking with at the time and are a reflection of my identity as an educator, researcher, and designer. To situate rules in the experience of the students, a set of *rules for mapmaking* was documented on the walls towards the end of our time in the space.

The Unruliness of Rules

I use the term *unruliness* in reference to the origin of the word *rule*, meaning *straight stick* (Daston, 2022). In this section, I hope to show how the rules, rather than keeping me straight, in practice encouraged me off a familiar path.

The rules for researching are about creating conditions for things to happen, rather than instructing me to discover or find answers. They are guidelines that embody a particular orientation and acknowledge personal habits that might get in the way of that orientation. The rules emerged as a personalised, situated, and informed set of instructions that, if followed, promised the research would "go well." To show how the use of the rules played out in practice, I explore three distinct examples. First, I consider how they acted as *ground rules* in the setting up of the space; next, I use an example of how they worked as an *emergency protocol* to rescue a situation; finally, I show how they extended beyond their instruction as *expert guidance* that helped me develop my research craft.

Ground Rules

As in sport, *ground rules* are about setting up the field for play. Early in the planning stages of my PhD, I came up hard against the question of methodology: What methods will you use? What data will you gather? How will you analyse that data? What are the ethical implications? Taking a post-qualitative stance and speculative approach to researching means that many of these questions are not useful ones; they assume a sequence of discrete activities that I wanted to disrupt (St. Pierre, 2021). My methodological conundrum was about how I could create conditions in the research event that permitted engagement with, and exploration of, the dynamic complexity of everyday life using mapmaking. To help me do that, certain rules became important as *ground rules.*

An important ground rule was that everything that came into the space, or was created in the space, should remain in play. **Rule #3**, **Keep everything in play**, is a thin rule (Daston, 2022), which means there is no room for deliberation on the part of the researcher: *everything* means *everything*. This ground rule is a way of acknowledging that everything is potentially of interest to the research event. It is a principle of practice that resonates with the art studio wall as an eclectic space for objects of interest to gather without knowing how and when they might spark an idea. An example is the expanded studio practice of the artist Olafur Eliasson (Godfrey & Eliasson, 2019) where ideas emerge from serendipitous associations between materials collated over time. Or in the practice of *open studios* that is familiar to my work with the design team at City ID (www.cityid.com), where the iterations of design remain visibly displayed around the room. Or in the social sciences where, for example, Chappell et al. (2022) make use of the materiality of data and the possibilities of digital whiteboard spaces to bring into relation theory, observations and the trajectory of ideas as a transformational practice. With its origins in the concept of multiplicity, this ground rule is a way of evoking a commitment to immanence because it is a commitment to the concept of relational space, of acknowledging the multiplicity of meanings, memories and relationships that came with each of the objects we created, and their potential to shape the nature of the moment we found ourselves in (Harvey, 2019).

> It is more than halfway through the afternoon workshop. We are using the technique of redaction to black out the words of a text we all have — a section of transcript from a recording made the previous week when we talked about our personal geographies map. Some of us still have the black markers in our hands, the width of the nib perfectly matched to the height of the letters, so words disappear in one stroke of the hand. As we begin to lift the page of redacted text, we see marks have been made by the ink soaking through the paper to the sheet below. A pattern of black marks runs in horizontal lines across the page. One of the students comments: "it's like The Matrix."[3]
>
> I think about how these pages of residue ink might be discarded; it's a sign that the paper is of inferior quality, its weight (thickness) is not fit for the purpose of using permanent markers. The paper is spoiled.
>
> But instead we must keep the sheets; the pages along with their ghost-like patterns are added to the wall. As we all watch them being displayed, another student observes: "That's what dyslexia feels like." This moment seems to be one that moves us all. I think about the possibility of these unreadable marks expressing something different for each of us. It is a glimpse of something that is in the process of making itself known. Sometime later, I notice the comment "lost words" has been added to the wall.

MacLure (2013a, p. 229) talks about "lively excess" as something that escapes representation. She draws attention to a form of knowing that is lost when we are limited to conventional structures and systems of communication based on language. Art practices—painting, drawing, performance, dance, sculpture—offer ways of mobilising these different kinds of

[3] The Matrix is a 1999 science fiction film that uses a visual effect of vertically falling computer code that, on first glance, appears as a fragmented pattern.

knowing, acting as a "language that speaks before words" (Deleuze, 1994/1968, p. 10).

The ground rule of keeping everything in play ensured that the artefacts that came into relation with each other were not only those that we recognised as relevant to the inquiry. The marks made by the ink bleeding through the paper were folded into the potential of the space. A visitor showed me a photograph on their phone. I printed it and added it to the wall. A student started a map again; the false start was not discarded. Research questions became rephrased or replaced as we moved through the six weeks; the visible crossing out and adding of words was evidence of this process.

What the ground rule of keeping everything in play does is encourage an acceptance of the messiness of research. It helps steer away from any instinct to make sense of what is happening and make any judgments about what is interesting or "good." It allows something that is created in the early stages to come into connection with an artefact produced much later, for unanticipated possibilities to emerge at any time, not only in the moment of creation or within a preconceived, defined timeframe for analytical thinking. The thinness of the rule—its lack of flexibility—takes the responsibility for determining what matters in an event out of the control of the research and researcher.

Emergency Protocol

As I write this chapter, I have beside me a small leather-bound book, no bigger than the size of my hand. Embossed on the cover is the British Railways marque and the words RULE BOOK 1950. I read how many of these rules correspond to a particular situation: "Should the following happen… the rule is…" These rules are protocols—actions that will rescue a situation from potential danger.

> It is quiet in the research space. The students have left, and I am standing in front of the wall where what was produced in the workshop earlier that afternoon is taped up. The maps I am looking at are different from the other maps around them. There are headings and subheadings; there is structure and order; there is a regular rhythm; and, when I half close my eyes, there is a uniformity. These maps are marked with headings: Year 8, Year 9, Year 10, Year 11, Year 12, Year 13, The Future.

These are timelines.

As I stand in the silence of the space, I feel anxious. I recall a conversation from the workshop in which one of the students told me how they feel about their studies. swirling around in the conversation are anxieties about studying during the Covid period, relationships with teachers, memories of things they enjoyed when they were younger, and hopes for the future. "Where is that in the map?" I ask. We look at the paper. The linear sequence of time doesn't allow space for what is being felt.

Now I am hearing a voice in my head. A member of the panel that reviewed my progression from proposal to PhD asks: "What are you going to do if they lapse into tracing?" Tracing is a concept associated with the rhizome (Deleuze & Guattari, 2013/1988). A tracing describes how sets of relationships conform to pre-conceived structures. Here in the research space, the shape of the maps follows a convention; it is a visual representation of time as linear, sequential and uniform. The timeline puts at risk any possibility of the new; as a student observes, this is the "official" story of secondary and post-16 education. The tight confinement to a pre-given structure results in a kind of repetition of the familiar and obscures the sensed, felt, remembered and hoped.

I think about my response to the question: "What are you going to do if they lapse into tracing?" I consult the rules for researching on the wall opposite. **Rule #9 Fall off edges** dares me to move beyond this familiar structure of the timeline. **Rule #8 Stir things up** asks me to observe a principle of intentional disruption. I write the map instruction for the next session: Collapse time. Cut up a timeline. Make fragments. Remake it. Consider relationships.

Whilst encouraged that cutting up and remaking the maps using a technique of collage might rescue the tracings by surfacing something that, as yet, had been hidden from view, I also felt this move was taking, quite literally, the notion of "putting the tracing back on the map" (Deleuze & Guattari, 2013/1988, p. 14). The rhizome chapter (Deleuze & Guattari, 2013/1988) helps me think of disruption as a productive event rather than a destructive force that determines boundaries or that separates elements. Transformation happens, not by extension, but through breaking, restarting, and taking new directions. It is through these ruptures that we extend our territory; it is how we express and understand our world and develop our sense of self. By using only the timeline as a resource to produce an alternative map, we were testing the possibility of disrupting the hierarchy in order to augment the relational properties—and therefore the possibilities—of the map (Corner, 1999).

The rule that came to matter as emergency protocol—**Stir things up**—is *thick*. Thick rules account for a range of different situations and require some interpretation as to how they might be applied (Daston, 2022); the rule is to stir things up, but *how* to stir things up is open for debate. In contrast, **Rule #9 Fall off edges** is *thin;* it instructs us to move beyond a place of safety. The *edge* belongs to a static, predictable world; a world where time moves from left to right along a marked path. The rule instructed me to break beyond the familiar framing of time, to rescue life from it. It was the move that is needed to make "drastic changes to our familiar mind-sets and established values" (Braidotti, 2019, p. 3). The edge—which felt in this moment of mapmaking like a dead end—was transformed; it became a point from which to construct an alternative story. These contrasting characteristics of rules matter. Thick rules are open to how the situation could be reframed; thin rules help to mobilise rapid action.

The removal of the "official" story of a linear, chronological process of change activated a more personal, relevant, and experiential expression of the students' situation. The mapmaking for collapsing time produced what I came to see as the "thick present, the moment of encounter between past and future" (Facer, 2016, p. 71). The two mapmaking sessions marked a shift from the official to the unofficial story of students' lives, as a student observes, from the factual *what* of their lives to a contemplation of the *why*.

Expert Guidance

This final example explores how the rules operated beyond their instruction. In The Craftsman (Sennett, 2008), Richard Sennett signals the limitations of written instruction in the mastery of crafts. He says: "Language struggles with depicting physical action, and nowhere is this struggle more evident than in language that tells us what to do" (p. 179). In traditional crafts, the apprentice—rather than reading instruction—learns from the master by watching, listening, and mimicking actions. But this kind of learning might not be accessible to all who wish to develop their craft; at the start of my studies, opportunities for me to be an apprentice to a master speculative researcher were not forthcoming. Sennett suggests: "Rather than get rid of print, the challenge is to make written instructions communicate to create expressive instructions" (p. 179).

One way that expressive instructions can work is through recognising the connection between body and language. The use of verbs of bodily action in the rules—to stir, to fall, to make space, to use soft eyes—creates an image of how one might observe the rule. Sennett also highlights the use of metaphor as a means of producing an image or sensation that evokes the task at hand; he suggests that such instructions serve as language tools that connect the technical craft of doing and making to the imagination (Sennett, 2008).

It is the function of metaphor that I want to use in this final example of how my rules for researching performed in the endeavour to create conditions for possibility. To simplify how I show this, I will use the examples of **Rule #2 Don't get in the way** and **Rule #5 Use soft eyes**, although many of the other rules perform in a similar way to move action beyond the instruction itself.

As I started to set up the research space, I made the decision to minimise any ownership of the material artefacts in the room. This means maps that were produced were not named; it was not possible to know who had made comments on sticky notes, who had taken photographs or even who said the words that appeared in quotation marks—on the walls, there were direct quotations that did not carry a reference to their original source. Less a direct response to a particular rule or rules, flattening hierarchies and reducing boundaries in this way was more about creating the possibility of relation between objects as we worked and thought with the materials that were produced.

Earlier in the chapter, I mentioned my interest in Maggie MacLure's notion of wonder (2013b). The idea of wonder—of something in the space inviting us in—manifested in different ways in the space, for example: everyone who visited was invited to take a photograph of anything that moved them, to print it and add it to the wall; sticky notes were left around the room for people to comment on something that felt important; everyone was invited to move artefacts, placing one next to another that seemed to speak to it. These were instinctive responses that were made without explanation—a surrender to wonder. Could these moves have worked so freely if objects had carried authorship?

Authorship introduces hierarchies and power. Consider how the addition of a philosopher's name, or the name of a friend's ten-year-old daughter who made a map, would have changed our response to these objects.

Authorship also fixes the artefact in time and in sequence with other artefacts. Consider how we might have responded differently if we had insisted on a chronological order to objects that were still in a process of change. If a function of non-representational work is to be in the "midst of things" (Lorimer, 2015, p. 181), it feels important for the midst to be freed from the framings that might hinder its possibility of change. I recall a student choosing something on the wall that resonated with them. They told me: "It looked at me, and I said, 'yeah, I want you'". This possibility of connection suggests that the refusal to tie artefacts to a person, place, or time enabled them to move (us) more freely, creating a pedagogic condition that permitted the kind of encounter with difference that shifts our sense of self and the world around us (Verger, 2025; Verger et al., 2024).

Nonetheless, avoiding authorship of texts and images in the room was neither easy nor perfect. On the one hand, not naming the maps produced by the students felt like a positive step to respecting the students' right to anonymity, allowing them to decide at the end whether they wanted to be named or not in the research. Yet, it also presented the problem that, in writing up the research, their names might not appear, while other images and texts would be referenced. From a logistics perspective, I planned to return the original maps to the students at the end of the residency, so I came up with the idea of stamps with their initials—like the ones used in coffee shops for loyalty cards. This turned out to be a good move, as it emerged that the students took great pride in stamping their maps, but I sensed that the absence of full names permitted the materials to be more readily shared when, for example, they took sections of each other's maps to use in their own collage, or when they took multiple maps and layered them digitally to create a new kind of map—one where it was no longer possible to distinguish the individual parts.

It also felt uncomfortable not to conform to academic standards for referencing a quotation from a philosopher, theorist, or scholar—although I was sensitive to the barrier this could create for some and to the fixing of meaning that might result. I anticipated the need to reference materials in the formal writing up of the research and devised a miniature code that helped me trace the origin of various texts; a code that others would not find meaningful. This move gave me some comfort, yet I was unsettled that my own access to these codes would get in the way of the potential of the materials, emphasising the need to adhere to **Rule #5**: **Use soft eyes**. With partial success, the anonymity of the artefacts permitted connections that produced new ideas and concepts.

> I am in the space looking at two sticky notes that were moved the previous day so that they now sit next to each other on one of the walls. The first sticky note reads "not quite art." The other, "it's not not art." Both speak to the practice of mapmaking that covers the walls around them. Each of the sticky notes is written by different people who visit the space. The first is a 17-year-old student in Bristol, UK; the second is a Distinguished Professor of Education from Penn State, US. I know there is a space of 10 days between the creation of the two notes. The second was written in response to the first. I am enjoying the idea of a transatlantic, multigenerational conversation happening in multiple times and spaces between two people who never meet. The concept of conditions for possibility as being about an "in between" is strengthened by this exchange.

The two sticky notes helped me see how participatory mapmaking operates outside the genres of traditional mapping (Gerlach, 2014). Being both "not quite art" and "not not art" situates mapmaking at an intersection of modes. The notion of ***Intermedia*** helps out here. Intermedia is a term introduced to the Fluxus art movement in 1966 by co-founder Dick Higgins (2001). It denotes a practice that sits at the intersection of two or more modes of artistic expression. The term troubles the tradition of framing of artistic forms—such as painting, sculpture, music, or performance—as separate and insulated from one another; I draw a connection here with non-representational methodologies that challenge the privileging of single and more traditional modes of expression (Vannini, 2015). The concept of Intermedia is productive in how it helps surface "a between-ness or liminal state that creates new zones of participation, interaction, and understanding" (Miles & Springgay, 2020, p. 1011). The exchange of sticky notes produces the possibility of seeing these new zones of expression and participation. It matters in non-representational methodologies that objects are afforded equal footing to the human actors (Vannini, 2015). By affording these material objects our attention, we reduce the barriers that can obscure these productive relationships (Kotze et al., 2018). I want to suggest here that the lack of ownership displayed on the materials enabled the objects to gain that kind of attention.

As I mentioned, the principle of displaying materials without authorship did not result directly from a rule. To understand the relationship between the rules for researching and the possibility that emerges in this example, I return to Sennett and his concept of expressive instruction (Sennett, 2008). **Rule #2 Don't get in the way**, or **Rule #5 Use soft eyes**, uses metaphor. Both rules might be taken literally, but their power lies in their comparison between a recognisable action of the body and what it means to be a researcher in the space. Sennett suggests, following philosopher Donald Davidson, that metaphors are forms of language that are *expansive*. In these examples, the metaphor results in *more meaning* being produced than the instruction itself. As an example, to stir things up—based on the metaphor of animating a calm situation—evokes a practice of making trouble or deliberately complicating a situation. Sennett calls this form of language "metaphors as process" (Sennett, 2008, p. 192). Removing authorship of the materials was less a response to an explicit instruction than a response to a promise of a process that might be activated.

Conclusion

By exploring three situations in the research event, I hope to have brought to life how the rules for researching operated to produce the conditions for new possibilities. Daston (2022, p. 20) writes that the application of rules can be about "the optimal balance between predictability and spontaneity." She infers that predictability comes about through the successful application of rules and spontaneity through their absence. But the kind of spontaneity we need in order to challenge the frames and structures that limit genuine thought (Deleuze, 1994/1968) is not easily gained. These structures and frames often go unnoticed—they have become calcified through years of practice and the absence of challenge. What this means is that *not* having rules might not be enough to liberate us from the business-as-usual ways of organising the world and for new possibilities to arise. In this chapter, I have hoped to show how rules help in a number of ways: to set up an event in a particular way; to rescue us when things go wrong; to foster our expertise by enabling an image of what it feels like when something is "done well". The point I make in this chapter is that spontaneity and chance need rules too. I propose that rules should be among the tools that mediate the possible (Glăveanu, 2023). What endures is the challenge of making the right kind of rules.

Rules for Making Rules

Make thin rules to avoid the greatest dangers. *Thin* rules are designed to be more rigidly applied. There is less room for adaptation or for the researcher to use their discretion. Thin rules assume the world to be static and predictable. They can help create change in even the most difficult, stuck moments. To make thin rules, acknowledge your greatest fears and address them. Fear silence? Make a rule not to break it. Tendency to tidy up? Make a rule to resist categorisation.

Make thick rules that require you to negotiate how they might be observed. In contrast to thin rules, *thick* rules are more flexible in application. They are open to interpretation and are descriptive rather than prescriptive. Thick rules invite the researcher to use discretion; how the rule is applied takes into consideration the context and situation. Part of this negotiation is a judgement about whether it is, indeed, the rule that should be applied.

Don't be afraid of foraging for rules. Rules are all around us, and researchers should be encouraged to be playful. Applying a rule to a new situation can provoke unexpected results. Take, for example, the swimming pool rule: "Walk, don't run" or the road safety rule "Adjust your speed for different conditions", these foraged rules might be taken up and applied in the research space with productive results.

Don't try and imagine the situation you will need the rule in. Make it, and trust it will be the right rule. Rules can quickly become attached to a particular circumstance. Take the example above: the rule to "Adjust your speed" anticipates a car driver navigating a busy city safely. However, if we set aside that association and avoid anticipating its use, it can encourage a particular movement in space, the practice of writing, or the rhythm of a research project over a year. Trusting that the rule that has come up is the right one opens the possibility that it will apply in unexpected ways.

Create an artefact of your rules. Spend time with them and make them special. Slowing down the writing process lets you take the time needed to get the wording just right. By spending time with the words—for example, by writing the rules out by hand, or using a stencil to draw each letter—they become more familiar, more comfortable. In the case of the rules for researching above, I began by creating a poster version, something legible from where I stood in the room. Later, I printed the rules on cards using letterpress—a slow process of relief printing that indents the rules into the surface of the paper. The crafting of these palm-sized cards took time and gave the rules some kind of permanence—it was a commitment.

Keep rules visible. However, you choose to record your rules, keep them visible at all times. The visibility of the rules is essential to the health and safety of the research because the situations that require you to refer to ground rules, emergency protocol, or expert guidance can show up in the process of researching at any time.

Be vulnerable when you make the rules; they can rescue you from your own limitations. To do their best work, rules should be deeply personal. Ask yourself, what could go wrong? Acknowledging your own edges, recognising ingrained habits, noticing your worst fears, and generating rules in response to these mean your rules will work hardest for you and in the context of your research.

Enter all rules into a process of negotiation and continue to do so as you observe the rule. Allow and expect the rules to generate internal and external dialogue and reflexivity. Rules, whether thin or thick, require negotiation among multiple identities, concepts, theories, and sensibilities and invite us to engage with our deepest assumptions and beliefs. This process is an expression of the active role of the researcher, participants, and others in the research.

Don't worry if you don't follow the rules. A rule defines a trajectory; it is part of a process of change. The practice of rules is a learning process that leads to change. Ignoring or breaking the rules must remain a possibility in this process and offer valuable learning.

Expect them to change. Rules, like life, need to keep moving.

References

Alvermann, D. (2000). Researching libraries, literacies, and lives: A rhizoanalysis. In E. St. Pierre & W. Pillow (Eds.), *Working the ruins, feminist poststructural theory and methods in education* (pp. 114–129). Routledge.

Boyd, C. P., & Edwardes, C. (Eds.) (2019). *Non-representational theory and the creative arts.* Palgrave Macmillan. https://www.dawsonera.com:443/abstract/9789811357497

Braidotti, R. (2019). Affirmative ethics and generative life. *Deleuze and Guattari Studies, 13*(4), 463–481. https://doi.org/10.3366/dlgs.2019.0373

Chappell, K. (2022). Researching posthumanizing creativity: Expanding, shifting, and disrupting. *Qualitative Inquiry, 28*(5), 496–506. https://doi.org/10.1177/10778004211065802

Corner, J. (1999). The agency of mapping: Speculation, critique and invention. In D. E. Cosgrove (Ed.), *Mappings* (pp. 213–252). Reaktion Books.

Craft, A. (2015). Possibility thinking: From what is to what might be. In R. Wegerif, L. Li, & J. C. Kaufman (Eds.), *The Routledge international handbook of research on teaching thinking* (pp. 153–167). Routledge.

Cremin, T., & Chappell, K. (2019). Creative pedagogies: A systematic review. *Research Papers in Education, 36*(3), 299–331. https://doi.org/10.1080/02671522.2019.1677757

Daston, L. (2022). *Rules: A short history of what we live by*. Princeton University Press. https://muse.jhu.edu/book/101548/

Deleuze, G. (1994). Difference and repetition (P. Patton, Trans.). Columbia University Press. (Original work published 1968).

Deleuze, G. (1998). What children say (D. Smith & M. Greco, Trans.). In *Essays critical and clinical*. Verso (1993).

Deleuze, G., & Guattari, F. (1994). *What is philosophy?* (H. Tomlinson & G. Burchill, Trans.). Verso. (Original work published 1991).

Deleuze, G., & Guattari, F. (2013). *A thousand plateaus*. Bloomsbury. (Original work published 1988).

Dewey, J. (1963). *Experience and education*. Collier Books.

Facer, K. (2016). Using the future in education: Creating space for openness, hope and novelty. In H. E. Lees & N. Noddings (Eds.), *The Palgrave international handbook of alternative education* (pp. 63–78). Palgrave Macmillan. https://doi.org/10.1057/978-1-137-41291-1

Firebaugh, G. (2008). *Seven rules for social research*. Princeton University Press.

Gerlach, J. (2014). Lines, contours and legends: Coordinates for vernacular mapping. *Progress in Human Geography, 38*(1), 22–39. https://doi.org/10.1177/0309132513490594

Glăveanu, V. P. (2023). Possibility studies: A manifesto. *Possibility Studies & Society, 1*(1–2), 3–8. https://doi.org/10.1177/27538699221127580

Godfrey, M., & Eliasson, O. (2019). *Olafur Eliasson: In real life*. Tate Publishing.

Hand, H. (2024). *Stirring up new ideas: Mapping as a tool in the practice of creative thinking*. University of Bristol.

Harvey, D. (2019). Space as a key word. In *Spaces of global capitalism: A theory of uneven geographical development* (pp. 78–95). Verso. http://ebookcentral.proquest.com/lib/bristol/detail.action?docID=5651421

Higgins, D. (2001). Intermedia. *Leonardo, 34*(1), 49–54.

Jackson, A. Y., & Mazzei, L. A. (2018). Thinking with theory. In *The Sage handbook of qualitative research* (5th ed., pp. 717–737). SAGE Publications Inc..

Kotze, R., Perold-Bull, K., Costandius, E., & Alexander, N. (2018). Posthumanist and new materialist perspectives as navigational tool in art education research: A diffractive analysis. *Critical Studies in Teaching and Learning, 6*(1), 10.14426/cristal.v6i1.124.

Lanier, V. (1965). An interview with Sister Mary Corita. *Art Education, 18*(2), 10–14. https://doi.org/10.2307/3190674

Lave, J., & Wenger, E. (1991). *Situated learning: Legitimate peripheral participation*. Cambridge University Press. http://catdir.loc.gov/catdir/enhancements/fy0702/91010450-t.html

Leander, K. M., & Rowe, D. W. (2006). Mapping literacy spaces in motion: A rhizomatic analysis of a classroom literacy performance. *Reading Research Quarterly, 41*(4), 428–460. https://doi.org/10.1598/rrq.41.4.2

Lorimer, H. (2015). Non-representational theory and me too. In P. Vannini (Ed.), *Non-representational methodologies: Re-envisioning research* (pp. 178–187). Routledge.

MacLure, M. (2013a). Researching without representation? Language and materiality in post-qualitative methodology. *International Journal of Qualitative Studies in Education, 26*(6), 658–667. https://doi.org/10.1080/09518398.2013.788755

MacLure, M. (2013b). The wonder of data. *Cultural Studies ↔ Critical Methodologies, 13*(4), 228–232. https://doi.org/10.1177/1532708613487863

Makelburg, D. (1967). *Corita Kent's rules and hints for students and teachers.* Corita Art Center. https://www.corita.org/tenrules

Manning, E. (2015). Against method. In P. Vannini (Ed.), *Non-representational methodologies: Re-envisioning research* (pp. 52–71). Routledge.

Marshall, K., & Loydell, R. (2015). Control and surrender: Eno remixed—Collaboration and oblique strategies. In S. Albiez & D. Pattie (Eds.), *Brian Eno: Oblique music* (pp. 181–192). Bloomsbury.

Massey, D. (2005). *For space.* SAGE Publications.

Mazzei, L. A., & Jackson, A. Y. (2023). Inquiry as unthought: The emergence of thinking otherwise. *Qualitative Inquiry, 29*(1), 168–178. https://doi.org/10.1177/10778004221096854

Miles, J., & Springgay, S. (2020). The indeterminate influence of Fluxus on contemporary curriculum and pedagogy. *International Journal of Qualitative Studies in Education, 33*(10), 1007–1021.

Rossetto, T. (2020). *Object-orientated cartography: Maps as things.* Routledge.

Savransky, M., & Stengers, I. (2018). Relearning the art of paying attention: A conversation. *Substance, 47*(1), 130–145.

Savransky, M., Wilkie, A., & Rosengarten, M. (2017). Speculative research: The lure of possible futures. In A. Wilkie, M. Savransky, & M. Rosengarten (Eds.), *The lure of possible futures: On speculative research* (pp. 1–17). Taylor & Francis Group. http://ebookcentral.proquest.com/lib/bristol/detail.action?docID=4809749

Sennett, R. (2008). *The craftsman.* Penguin Group.

Shaughnessy, A. (2021, May 24). Adrian Shaughnessy on Ken Garland, a "disruptive and questioning spirit". *It's Nice That.* https://www.itsnicethat.com/news/obituary-ken-garland-adrian-shaughnessy-graphic-design-240521

Springgay, S., & Truman, S. E. (2017). On the need for methods beyond proceduralism: Speculative middles, (in)tensions, and response-ability in research.

Qualitative Inquiry, 24(3), 203–214. https://doi.org/10.1177/1077800417704464

St. Pierre, E. A. (2021). Why post qualitative inquiry? *Qualitative Inquiry, 27*(2), 163–166. https://doi.org/10.1177/1077800420931142

Thrift, N. J. (2008). *Non-representational theory: Space, politics, affect*. Routledge. https://www.dawsonera.com:443/abstract/9780203946565.

Tromp, C. (2023). Integrated constraints in creativity: Foundations for a unifying model. *Review of General Psychology, 27*(1), 41–61. https://doi.org/10.1177/10892680211060027

Truman, S. E., Loveless, N., Manning, E., Myers, N., & Springgay, S. (2020). The intimacies of doing research-creation. Sarah E. Truman in conversation with Natalie Loveless, Erin Manning, Natasha Myers, and Stephanie Springgay. In N. Loveless (Ed.), *Knowings and knots: Methodologies and ecologies in research-creation* (pp. 221–250). The University of Alberta Press.

Vagg, J. (2022). Experiencing-with data: Exploring posthuman creativity through rhizomatic empathy. *Qualitative Inquiry, 28*(5), 541–551. https://doi.org/10.1177/10778004211069696

Vannini, P. (2015). *Non-representational methodologies: Re-envisioning research*, P. Vannini, Ed. Routledge.

Verger, N. B. (2025). *The aesthetics of the possible*. Possibility Studies & Society. https://doi.org/10.1177/27538699251318068

Verger, N. B., Duymedjian, R., Wegener, C., & Glăveanu, V. P. (2024). Creative preservation: A framework of creativity in support of degrowth. *Review of General Psychology, 28*(3), 268–281.

Watershed. (2023, January 20). *Stirring up new ideas: Unreliable mapmaking* [Video]. YouTube.

CHAPTER 7

Dances Yet to Come: The Possibilities of Performance that Never Was—Or Only Ever Lived in Our Imaginations

Vicky Hunter

Introduction

This chapter explores themes of imagining and being "otherwise" (Grosz, 2011, p. 87), speculative fabulation (Haraway, 2016), and corporeal projection in relation to specific pedagogical challenges posed during the COVID-19 pandemic lockdowns, particularly the cessation of live performance and assessment work. It articulates a creative response to the challenge of preparing Higher Education Masters-level dance students for the development, presentation, and assessment of choreographic work that may never be performed. Drawing on ideas of embodied and poetic imagining and creative dreaming (Bachelard, 1994, Hunter, 2021, Le Guin, 1976), it responds to the following questions:

V. Hunter (✉)
Bath Spa University, Bath, UK
e-mail: v.hunter@bathspa.ac.uk

N. B. Verger, D. Mattson (eds.), *Possibility Studies and Societal Futures*, Palgrave Advances in Possibility Studies,
https://doi.org/10.1007/978-3-032-26379-7_7

- How might we reflect on imagined or possible dance-making processes?
- How might we evaluate the efficacy of possible strategies and approaches "applied" within the making of a dance that is yet to come?
- How does this form of future-thinking shape and inform pedagogy in this context?

This writing emerged in response to pedagogical challenges arising from the cessation of face-to-face teaching and supervision of performances and live assessments, which I encountered whilst leading the MA Choreography and Professional Practices programme at the University of Chichester, UK. It responds to the issue of negotiating students' aspirations to present live choreographed dance work for assessment with the reality of developing and re-imagining dance performance work as an entity that may never be performed. The chapter also encompasses subjective reflections on my own positionality as a creative practitioner-researcher, individual, and pedagogue situated within the early stages of UK lockdown constraints, and articulates some of my responses to navigating this newfound territory.

The task of imagining a dance or performance work that may never be performed is not uncommon in creative arts practices (and related fields such as architecture and design). Architects, for example, regularly make speculative models of buildings that are never built. These spectral traces of virtual buildings without tangible form are often depicted as situated in real-world locations, yet factors such as financial or planning constraints often prevent their completion. In the fields of dance and performance, constant cycles of arts funding and grant applications, central to portfolio working, render practitioners familiar with this mode of speculation and possibility as they aspire to produce the as-yet-unformed germ of an idea through hypothetical projection and imagination.

When working with students, however, the shift from a tangible product to imagined and speculative processes raises both challenges and opportunities, inviting reflection on how we might capture nuanced engagements with arts-based Practice-as-Research whilst edging towards an outcome that might not, as yet, be realised. Differentiating Practice-as-Research from artistic practice, Robin Nelson (2022, p. 22) describes a form of research that is "undertaken through a practice (praxis) and that the processes or products might constitute the primary evidence of the research"; furthermore, he suggests that researchers working in this mode might "attend differently to their work in presenting it *as* research" (emphasis in the original). Writing about creative alternatives to the UK's prevalent "anticipation" educational regimes, Sarah Amsler and Keri Facer (2017, p. 19) critique systems that gesture towards futures yet to come, yet maintain the *status quo* despite the potential to propose alternatives. This, they argue, leads to scenarios of "hopelessness" in which the possibilities for change and the imagining of alternative futures are obscured. They ask:

> How can people conceptualise the relationship between education and the future in such a way that learning not only enables collective survival in the present ... but facilitates the invention of new capacities which enlarge possibilities for yet- unimaginable alternative futures to emerge?

Informed by their thinking and writing, my reflection on developing dances-yet-to-come centres on the pedagogic potential of bringing the future into students' processes of thinking, being, and doing in the present moment, thereby fostering a mode of present-futuring in which potential scenarios of becoming are modelled in the here and now. In this scenario, it may be possible to imagine and create a dance-yet-to-come that feels and materialises as present and affective, thereby invoking creative, tangible, and hopeful responses to possible trajectories of performance-making in the here and now.

This hypothesis raises the following questions: How might the imaginary and speculative modelling of draft movement phrases, and the re-configuring of scenographic and presentation formats, function as creative strategies through which some degree of tangibility might emerge? How might material developed from this way of working enable us to embody, feel, and sense our way towards sharing this material in the future with other bodies on stage? And how might future thinking applied in this context shift and inform pedagogical approaches to choreographic teaching, working, and learning?

This reflection is further informed by the context of the wider edited collection in which this chapter is situated, and by its concern with the notion of affordances and how the affordances of a given situation or circumstance shape behaviours and actions. Writing on relationships between organisms and their environment, J.J. Gibson (1979, p. 68) offers an often-cited observation that feels pertinent to my enquiry when he notes, "the affordances of the environment are what it offers ... what it provides or furnishes, either for good or for ill". Whilst Gibson's writing in this example is concerned with terrain, objects and the materiality of physical environments and what these entities collectively afford the inhabitant or experiencer, I am extending his ideas here to encompass the affordances of institutional structures and pedagogical protocols that, operating in relation, construct and shape working environments with specific valences and affordances. The perception and embodiment of these affordances, and how they shift and become affected/affective during times of (pandemic) crisis, reveal potential for thinking and operating differently in ways that both constrain and release practitioners in specific environments (in this case, Higher Education).

Ideas and perspectives drawn from feminist new materialism and post-qualitative enquiry, as articulated by scholars including Elizabeth Grosz, Donna Haraway, and Karen Barad, who champion alternative ways of knowing and knowledge production from an ethico-ontological perspective, also inform this reflection on research and pedagogy (see also, Burnard et al., Chap. 5, this volume).

To those committed to social, political (and educational) change, the feminist scholar Elizabeth Grosz (2011, p. 87) urges us "to think otherwise, to become otherwise". Eschewing traditional epistemological quests for truth, logic and representation, she champions the potential for feminist theory and new feminist ontologies to "make us become other than ourselves, to make us unrecognisable", and, through engagement with the entangled complexity of the world beyond our own sense of subjectivity, she encourages us to embrace "what we don't control or understand in order to expand, not confirm what we know, what we are, what we feel" (Grosz, 2011, p. 87). This call for thinking and imagining otherwise chimes with my own attempts, articulated here, to encourage students' processes of creative dreaming and embodied, poetic imagining, the possibilities of which, I propose, can be multiple and liberating.

Making a Dance that May Never Be Performed

In developing a contingency plan that postponed the presentation of Master's-level practice-as-research dissertation work from June 2020 to an unknown future date, my colleagues and I faced a dilemma about how best to prepare students and guide them through a practice-as-research process that, for the majority of the cohort, would involve "making" a dance work that might never be performed. Practical solutions to this dilemma were workable, namely the provision of alternative submission formats and the possibility of extending deadlines, with protocols for their implementation in place. Fortunately, our assessment criteria and learning objectives were sufficiently flexible to encompass a range of alternative assessment formats, including essays, lecture demonstrations, creative portfolios, and remote presentations, as required.

On a practical level, therefore, a workable (but not ideal) scenario was established in which creative practice-as-research processes could be captured and articulated as fully as possible, with guidance and suggestions provided. However, at a pedagogical and philosophical level, the potential challenges and opportunities posed by this shift prompted my own period of reflection on how we capture nuanced engagements with practice-as-research processes whilst edging towards an outcome that may never exist. In this writing and reflection, I am mindful of the complexity of the pandemic and the existential crisis it invoked, which, for many, resulted in severe illness, bereavement and trauma. Alongside this, I am interested in further examining the ongoing legacy of the extraordinary spatial, embodied, and imaginative experiences encountered during this period, and the reconfigurations (both corporeal and organisational/institutional) they invoked. In the rush to establish the post-pandemic new normal, many of these deeply felt experiences and encounters were easily subsumed within the ongoingness of everyday life, yet the legacy of adaptive, creative working practices and alternative modes of thinking, being, and relating to others remains.

Whilst Practice-as-Research (PaR), by its very definition as a mode of research, encompasses process as an intrinsic component of the methodology, it nevertheless leads practitioners in most cases towards a tangible outcome (see Robin Nelson, 2006; Christine White, 2002; Melissa Trimingham's, 2002 writing on Practice-as-Research). Projecting towards a choreographic PaR process involves imagining the germ of an idea as an iterative creative process in which the accumulation of thoughts, ideas, materials and movement phrases creates a "floating nebula of ideas and possibilities in the mind that might at some point solidify into a clearer project that moves from the back to the front of the creator's mind" (Claxton, 2006, p. 63). In the new pandemic context, I found myself pondering how students might reflect on and articulate further their engagement with the preparation and planning of this process, the imagining of the process, and subsequent reflection on this envisioning. In a sense, they were preparing for a virtual outcome—not a technological virtuality but an imagined one. The guideline document submitted to the students to help them navigate this new terrain is presented below:

MA Dance Dissertation

Contingency Guidance April 2020

Please read this carefully:

Dear MA students—in light of the uncertain climate caused as a result of the University's adherence to the issued guidelines emerging from the UK Government's response to the COVID-19 pandemic—we may need to prepare for a dance/performance that may or may not happen.

As we approach July/August, maybe the picture will become clearer, but in the meantime you all need to prepare for an alternative assessment format.

As each project is different, this will depend on the nature of the work; some may choose, for example, to switch to 100% written at this stage or to increase the written component of the overall submission. Please remember the submission is marked as a whole project and given one grade.

For some of you, an alternative portfolio or lecture demonstration (recorded or presented remotely) might appeal as an alternative. This might be a useful alternative/mode of assessment for International students, for example.

What you all MUST remember is that this is a Practice-as-Research project and you can gather evidence of your research, thinking and planning from this point onwards—this is NOT an assessment of your ability to arrive in a studio (when we are allowed back in) and produce a dance or performance work in a week or so.

PLEASE liaise with your **dissertation supervisor** regarding your progress and plan for the project moving forward.

So, how to evidence research-led, choreographic planning/thinking and research becomes the new challenge.

You can start gathering evidence of this in a number of ways and can start collating materials now that might be relevant (this list is for guidance only and is not exhaustive, adapt and select as required):

- A detailed project plan outlining research question, theoretical influences, practical influences, methodology etc.
- Reflection on reading/materials/choreographer interviews that inform your approach to developing the work—this could be in short, reflexive pieces of writing presented in response to/alongside the original material (e.g. links to Pina Bausch's work or to an article in *Choreographic Practices*, *Dance Research* or *Animated* journal, for example); these reflections could also take the form of short video presentations in which you speak to the camera also.
- Video footage/photographs/images of developing material—this could be performed by the student/choreographer and discussed in relation to how this material would be developed/embellished through working with other dancers. It could include key motifs or the exploration of particular scores, for example.
- Choreographic scores and tasks—these can be very illuminating in terms of articulating process—what tasks/scores might be employed in the studio— what might their intended outcomes be and why were they designed in this way?) Steve Paxton's "Satisfyin Lover" (1967) is a great place to start—you have all worked through this score with me last semester.
- Examples and discussions of your plans for working with community groups/client groups/teaching groups (if this is part of your project): what tasks or sessions were you planning to lead and why—what might their imagined outcomes have been—which other practitioners (Deborah Hay, Alito Alessi, Anna Halprin, for example) run this type of session—what can you glean from their work?
- A clear choreographic/performance "storyboard" identifying key scene/section of the proposed work—these storyboard sections can also indicate lighting/sound effects or any scripted material.
- Floor plans (using Laban's method preferably)—to be included in the storyboarding of this imagined work—this evidences an understanding of choreographic design and thinking—and should be accompanied by reflection on these choices with reference to "classic" choreographic texts such as Blom and Chaplin, Preston-Dunlop, Jenn Joy, etc.

- Evaluation and reflection material—there should be evidence in the final portfolio/lecture demonstration presentation of reflection on this process so far—this will focus more on the research design at this stage (if the work does not get produced) and should reflect on how one thing has led to another—i.e. how did watching the work of Alexander Ekman then lead to an exploration (real or imaginary) of props and scenography, how did that then lead to a reading of Rachel Hann's work on stage design and what impact did these connections/journeyings through practice and theory have on your own thinking, doing and imagining?

These guidelines can be adapted to suit most projects—be creative in gathering your portfolio of evidence—the online *Journal for Artistic Research* provides some great examples of how to do this. If writing is not your thing, then record short audio or video reflections on your engagement with source materials, draw diagrams, paint pictures—whatever might help you show how you have been "making" this work or "leading" these sessions or "performing" in this work in your imagination.

ARTISTS DO THIS ALL THE TIME—we as a staff base have a great deal of experience of planning for and imagining dance and performance works and arts projects that NEVER get funded or performed, yet we have to convincingly tell others about them in advance!

IF the studios and the campus are open again in July/August and you can come back and start making this work for September performances, then great—this will be the most organised and well-researched work ever; if not then we will work with you and use our imaginations to assess the portfolio material and imagine a performance outcome that never was—or only ever lived in our imaginations—together.

As we shifted to online lessons, conventional, transactional modes of educational delivery in which knowledge and information are imparted, notes are taken, and evidence of learning is presented through discussion and ultimately the submission of assessment material were challenged during this period of lockdown teaching, as the whole person's circumstances began to shape and inform learning experiences in both overt and subtle ways, and a bleed between the domestic, educational, and professional learning spheres began to take place.

Teaching students in their domestic spaces revealed disparities of class and privilege, for example, as some struggled to find time on shared laptops or as younger siblings invaded the learning space. Students in dressing gowns, in their bedrooms, cooking meals, or in their gardens on sunny days fed into and shaped an approach to teaching and facilitating that morphed and shaped itself to the students' capacity for Zoom fatigue, information processing, sitting still, and sadness, as we moved through the rapidly dwindling academic year together potent with lost opportunities and diminishing hard-won dancing skills and physical fitness levels.

The tone of the final statement in the guidance for students presented here reflects a sentiment that runs throughout the document, and, on reflection, reads as a plea to keep going, keep thinking, reflecting and researching during an abruptly interrupted period of study. Reflecting on this document some years later, I can recall my own visceral sensations of urgency and consternation in communicating these thoughts and ideas, which manifested as guidelines for the students. I recognise that these concerns emanated from a deeply felt sense of care, compassion, and responsibility for students and their learning. They had trusted me and shared their creative ideas and aspirations with me this far on the programme of study, and I was acutely aware of the responsibility to hold space for these fledgling processes and, in partnership with the students, to ensure that these creative hopes, desires and aspirations did not wither and die, but rather morph, reassemble, and flourish in a differently formulated manner.

Aligning with the themes of Creative Preservation informing this book, this approach was very much concerned with preventing "the decay of existing materials and ideas by updating and adapting them, or re-expressing them in another way through the exploration of their affordances" (Verger et al., 2024, p. 270). The emotional resonance of this act remains a palpable trace as I recall these events in the moment of writing, and the associated shock, confusion, and panic that framed the early stages of the COVID-19 pandemic and shaped working practices and relationships in physical and emotional ways.

In the context of this practice-based research module, what we then became involved with was the cultivation and creative preservation of imagination and of speculation as vital strategies that support necessary processes of improvisation and adaptation that seemed particularly pressing in the uncertainty of the pandemic. This revealed an imperative to retain and foster awareness and openness to what-ifs, maybes, and journeying, enacting creative preservation as an "activity or process of keeping something which is valued alive, intact, or free from damage or decay" (Verger et al., 2024, p. 270), as opposed to arriving at outcomes, destinations, and a definitive end product.

Speculative Fabulation, Corporeal Projection, and Imagining Otherwise

The exploration of questions, propositions and proposals is central to choreographic thinking and movement research that involves "thinking-in" and "thinking-through movement" (Sheets-Johnston, 1969). Choreographic processes such as structured improvisation, movement development, and the forming of movement phrases and sequences engage movement researchers with speculative futuring through iterative cycles of questioning, as the performance studies scholar Gerko Egert (2019, p. 443) observes:

> Movement is brimming with questions. Movement asks questions, it reacts to questions, it can repeat and retain questions. The only thing it persistently avoids is answering questions. Choreography navigates by questions and with questions. It modulates movement's immanent force, its *questionness*: its *where-ness*, its *who-ness*, its *howness* (italics in the original).

Here, Egert captures the speculative nature of choreographic research and its attentiveness to the materiality of movement as an "immanent force", dynamic and iterative, feeding on and from recurrent processes of questioning, reflection, and counter-questioning. Aligned with this foregrounding of speculation and experimentation, the newly proposed assessment format presented to students and its associated implications for practice challenged my previous pedagogical approach to delivering assessment guidelines, information, and student supervision. The task in hand shifted towards a more discursive format in which possibility, suggestion, proposition, and speculation came to the fore as modes of thinking, doing, and practising. In previous writing, I have commented on the iterative modelling of the practice-as-research process as a form of spiralling towards an endpoint or end goal (Hunter, 2021).

Whilst iterative in nature, with a cyclical form of feedback loop in which the practitioner's process is fed by reflection on know-how, know-whether, and practitioner knowledge, there eventually comes a cut-off point for this process (usually an assessment point or performance outcome), wherein the process comes to an end or is temporarily suspended. Whilst a deadline for documenting the speculative imagining of this process remained, and without a final performance outcome, the ongoing potential to imagine and virtually create a dance that might live only in our imaginations could continue to develop and accrue creative input well into the future.

The type of creative imagining I advocated for at this time echoed through the *Imagined Theatres* online journal call for papers issued around the same time in April 2020 for a special issue "on emergency" led by Joshua Abrams and Daniel Sack, in which theoretical discussions of imagined theatres take place.[1] Responding to the COVID-19 emergency, they invited submissions that reflected on the effect of the crisis on non-performance-making as a form of collective endeavour, offering a platform in which performances could be shared and, collectively, participants could engage in "a virtual gathering where physical gatherings are no longer possible", the call continued:

> Theatres and arts institutions around the world have closed their doors and shuttered lights, interrupting premieres, runs, and rehearsal processes. Festivals are postponed, seasons suspended. The theatre is closed, yet the theatre keeps performing. Performances work their way out in our private thoughts or distanced conversations, in makeshift configurations across media. We imagine a release, a gathering together, an encounter with the sublime, a small act of repair. So, as communities fragment into socially distant parts, we propose a festival of imagined theatres alive to this state of emergency.
>
> In putting forth an "emergency" issue of *Imagined Theatres* we name both the occasion of our meeting and a mode of emergent becoming. The *Oxford English Dictionary* offers an obsolete understanding of emergency as "the process of issuing from concealment, confinement". So, too, we see this as an ongoing process: reviewing submissions as they arrive and issuing small groups of imaginary performances, week by week, from our confine-

[1] In a similar manner, in 2019, the Arts and Heritage cultural organisation opened a call for artists to respond to their blue skies museum initiative, which invited imaginary art installations and creative responses to UK heritage sites, none of which were (at the time) guaranteed to ever be realised: https://www.artsandheritage.org.uk/digital-projects/blue-sky-museums/

> ment. Around the world, people have been ordered to shelter at home, while leaving those without shelter or far removed from home exposed to a virulent atmosphere. Our private spaces, once concealed backstage, now appear as backdrops to remote conferences and classrooms, to virtual happy hours and dinner parties.

The notion of imagined performances proposed by Abrams and Sack (2020) extended their previous writing on Imagined Theatres, curated into an edited collection, in which they explore relationships between imaginary reflections and conceptions of performance work that inform and shape their manifestations in front of an audience. The extension of this work via the special issue in 2020 took on new poignancy in light of the pandemic and its impact on performance-making. Abrams and Sack's imagining of performances on a weekly basis during this time illustrates a form of creative imagining that harnesses the potential of performance-making and the mode of speculative fabulation I am advocating for in this writing.

Alongside the pedagogical challenges posed by the pandemic restrictions, my own creative endeavours as a site-specific researcher and dance-maker ground to a halt. During this time, I did not want to make any site responses, nor did I want to explore any scores or creative tasks. The irony of cajoling students to press on whilst my own creativity appeared to have evaporated was not lost on me.

I did, however, find myself moving around my house in a free-flowing, open-scored manner. Often, I would pause in a chair I rarely used before to view a familiar room from an unfamiliar angle. I lay down on a bed in a spare room and watched the patterns of clouds as they ambled past the skylight window. I often found myself on the floor, rolling, stretching, and pausing, reconnecting with my somatic body as an antidote to hours spent on Zoom calls with colleagues and students that had led to a frozen shoulder. Therefore, whilst any discernible outputs were not emerging from or associated with this activity, it provided space and time for me to connect with my own body and rediscover kinetic interests, pathways, and modes of moving that had previously been muted by an increasingly sedentary academic work and lifestyle.

The confines of the COVID-19 pandemic, therefore, led me to explore and re-evaluate previously familiar movement practices that might otherwise have been discarded or lost. By re-engaging with them, I acknowledged their inherent value, a perspective that aligns with Verger et al.'s

(2024, p. 274) discussion of creative preservation and its concern with "the continuity of what is already there, and variations of the modality of expression of this 'old' idea, behaviour, or artefact".

Initially resistant to any online movement sessions or technique classes, I gradually began to engage through gentle cajoling from dance friends and colleagues. To unfreeze my shoulder, I revisited the familiar movement patterns of the Cunningham technique, signed up for classes with Ellyn Hebron in the UK and with Janet Charleston and Jennifer Goggans in New York via the Merce Cunningham Trust's Instagram platform, and attended Gaga dance classes online. Gradually, the shoulder unfroze, my body came home, and I adapted to the new environment, its constraints, conditions, and opportunities. In response to these quiet shifts, I offered my own reflective writing to the students.

> I have been given the gift of time and, like most gifts received by me, I don't want to squander it or use it straight away. I want to sit with it and admire or engage with its gift-ness, its entity as a gift in its own right, a gift of time that does not demand an immediate plan but instead opens up a space to think, daydream, wonder,and just be. I am not keen or quick to open this gift and peel back its layers or to reflect on the conditions through which it was gifted to me. For the time being, I am content (or resigned) to be with the time given to me, to stay with its shape and contours, its ebb and flow, its form and density, to play with it in my hands and through my body, to sit with it for a while and see where and when it might unfold. (Vicky Hunter, note to students, May 2020).

Reflecting on the spatial constraints imposed by lockdown drew my attention to imaginary lines of flight that I describe as forms of "corporeal projection", helping to articulate what this embodied mode of imagining otherwise comprises. I turned to the writing of the spatial philosopher Gaston Bachelard, whose work informs my own research in site dance practice. Bachelard's writing both inspires and informs my engagement with space and place. He is a master of poetic imagining, conjuring images and alluding to, and crafting, what is not really there but might be if we delve a little deeper. Through suggestion, poetic inference, and at times bold proposition, Bachelard conjures words, worlds, and imaginaries by attending to sensation, smell, memory, embodied and phenomenological responses, and spatial and lived engagements with atmospheres, moods, and densities. His work also embraces a kinetic disposition towards phenomenological perception, with images and analogies that often mention journeying, moving, pausing by the fireside, wandering down corridors and hallways, crossing thresholds, and nesting and nestling into corner spaces.

My reading of Bachelard's writing on home spaces and universes, in particular, took on new resonance during the pandemic and associated periods of lockdown. In *The Poetics of Space* he explores the imaginative potential afforded by home spaces that enable a shift of perspective, an imaginative dreaming that sets in motion creative wanderings and discursive poetics. The home site in this scenario both protects us from the world whilst projecting us into it—and beyond:

> Thus, an immense cosmic house is a potential of every dream of houses. Winds radiate from its centre and gulls fly from its windows. A house that is as dynamic as this allows the poet to inhabit the universe. Or, to put it differently, the universe comes to inhabit his house. (Bachelard 1994, p. 51).

His suggestions regarding daydreaming as a oneiric form opened my awareness of and receptivity to the oneiric possibilities of dreaming otherwise. Whilst grounded in physical encounters with the architecture of home sites and informed by embodied memories of space and place (the odour of raisins, for example), Bachelard's invitation to engage with phenomenological encounters with lived space, informed by his reading of poetry and its invocation of embodied associations and sensorial affect, informed my sense of being grounded in my body in the physical lockdown space whilst extending a sense of perceptual awareness outwards into the world beyond. Through this process, my corporeal imagination manifested as a form of projection from which my body schema took flight, informed by his provocative prose, as I felt an embodied sense of (im)possible projection into the world beyond a situated location, flying over rooftops, lands, and territories, and hovering above situations and dwelling spaces.

Engaging with poetic and embodied processes of imagining otherwise fed back into my pedagogical approach to working with the students and enabled me to encourage their own articulations of creative dreaming, speculation, and ideation. Drawing on Bachelard's phenomenological approach alongside feminist new materialist and post-human approaches to creative writing and research creation (Truman & Springgay, 2015; McCormack, 2008), I developed short creative writing tasks and strategies for capturing reflections on Practice-as-Research that encouraged students to capture embodied experiences through metaphor, imagery, speculation, and attention to human-nonhuman relations. This approach aligns with the performance scholar Melissa Trimingham's (2002, p. 56) formative writing on Practice-as-Research and the role of the researcher-practitioner:

> ... it is the task of the researcher to translate such knowledge, however approximately, however unsatisfactorily, into analytical language, using metaphor, analogy, images, generously attempting to share with others the insight and understanding they have reached through their practice.

Together, we explored this new territory and worked to capture ongoing reflections on creativity, speculation, and "navigating the unknown" (Bannerman et al., 2006). This led to the inclusion of enhanced creative reflection on practice in the students' choreographic notebooks, the development of poetic prose and creative writing as modes for capturing creative ideas and experiments, and the inclusion of diagrams and drawings to illustrate often ineffable sensations, emotions, and experiences.

Unfettered by the logistics and problems of budget constraints, resourcing, and production schedules, the dance-maker's mind is free in this scenario to engage in creative dreaming as a form of speculative fabulation. As Ursula K. Le Guin (1976, p. 10) proposes, engaging with dreamscapes and the part of ourselves often excluded from everyday, rational lifeworlds encourages us to take dreams seriously and listen to dream-world scenarios in which "a great deal is [in] there, the things we most fear (and therefore deny) the things we most need (and therefore deny)". In her influential work Staying with the Trouble, Donna Haraway (2016, p. 203) describes speculative fabulation as a "mode of attention, a theory of history, and a practice of worlding".

Developed in critical dialogue with science fiction writing and storytelling, speculative fabulation works to problematise and propose alternatives to future-normative propositions of whiteness, ableism, heteronormativity, and masculinity often associated with the science fiction genre. Through speculation and creative articulations of envisioning a world otherwise, speculative fabulation proposes alternative visions of future worldings. As Sarah E. Truman (2018, p. 31) observes, "[w]hile rooted in everyday storytelling practices, speculative fabulation defamiliarizes, queers perception, and disrupts habitual ways of knowing". Speculative fabulation is therefore "not neutral", and our projecting towards the not-yet, the future yet to come (or not), is shaped by contemporary materialities, situated knowledges and contextual and historical legacies. In this way, "Speculative fabulations matter, and engender response(ability) for the futures we co-create through thinking, writing, and researching" (Truman, 2018 p. 33). To this, I add here moving, dancing, and choreographing otherwise.

Engaging with the Not-Yet

Many of the students, however, found it very difficult to imagine a possible future performance outcome that deviated from or developed beyond conventional modes of production and presentation. Perhaps this suggests a habitual mode of being and doing, shaped by a sense of "hegemonic futurity" proposed by Jeanne van Heeswijk, Maria Hlavajova and Rachael Rakes (2021, p. 12), when they discuss the possibility of "imagining and enlivening another chronopolitics". They continue:

> An alternative imaginary in its own right, the not-yet is a temporal category, though it is much more than a mere future projection. In its prospective orientation, however, it differs decidedly from the notion of the future offered under contemporary regimes. In western modernity, to put it simply, the future is a linear extension and a gradual expansion of the present, which itself arrived in the same fashion from the past. (van Heeswijk et al., 2021, p.12)

Drawing from van Heeswijk et al.'s (2021) perspective, engaging with the not-yet and imagining otherwise positions creative dreaming and speculative fabulation as a political act that challenges dominant "anticipatory educational regimes" that prevent "alternative knowledges and imaginaries" from thriving (Amsler & Facer, 2017, p. 11). Arguably, the affordances of these education regimes become embodied and embedded over time through repetition. For example, cyclical rounds of learning, assessment, and feedback reiterate institutional rules and protocols and facilitate measurable outcomes that in turn shape how teaching sessions and content are shaped, repeated, and reiterated. The structures of these regimes remain, for the most part, implicit, running in the background yet, returning to Gibson's (1979) writing, they afford those who work within them a perceived sense of "value" and "meaning" pertaining to the labour that is performed there and how it fits within the regime:

> This is a radical hypothesis, for it implies that the "values" and "meanings" of things in the environment can be directly perceived. Moreover, it would explain the sense in which values and meanings are external to the perceiver. (Gibson, 1979, p. 127).

Gesturing towards a dance that might never be made in this (pandemic) scenario opens up the potential for counter-hegemonic thinking that troubles the implicit values and meanings of the educational regime and its drivers for measurability, outcomes, and end products. Through drawing attention to alternative modes of being and doing that assign value to process, creativity, experimentation, and the material value of emergent "choreographic objects" (Neri & Respini, 2018), alternative affordances aligned with creative preservation begin to emerge:

> … creative preservation is a process that increases the value of things through exploring new uses, thereby amplifying the dignity of things, and thus our respect for them. The discovery of new affordances and possibilities for a given material enhances individuals' perceptions of the worthiness of the object which, instead of simply becoming disposable waste, is transformed into something new and useful, thus reinforcing the rationale for preserving the given material. (Verger et al., 2024, p. 276).

I am not sure whether my students ever fully reached a counter-hegemonic "aha moment" (to borrow from the choreographer and movement improvisation specialist Deborah Hay); however, as an anarchic provocation, the proposition outlined in this chapter offered liberatory potential for untethering from expectations of productivity, end goals, and finished products, and projecting towards the not-yet and what might never-be as an act of generative possibility.

Freed from the tyranny of completion, the imagined and speculative dance work regularly manifests for the choreographer as they explore its creative potential through various incarnations, with content morphing and changing in real time, in the here and now of the lived moment, rather than a postponed manifestation at a time as-yet-to-come. Encountered during the liminal space-time of the COVID-19 lockdown, this process was afforded time and space, and the various iterations of the dance work required attention from the students as they captured the import of things showing up for them on a regular basis, such as movement sequences, floor patterns, dynamic movement qualities, and scenographic components. The ability to spend time and dwell with these daily dances was contingent on the circumstances in which we found ourselves, removed from customary obligations, schedules, and activities. Students (and I) were able to attend to these imaginative and creative wanderings and mine their potential for reflection and insight into choreographic processes and practices.

Gradually, a noticeable shift also occurred in the student projects I was supervising as plans evolved and adapted, and creative solutions were found to problems posed by the uncertain conditions in which we found ourselves. One example was a site dance in small spaces project located in the domestic confines of a bedroom wardrobe. Another project explored dance and collectivity by considering what an online community might bring and mean to collective movement practice. Another explored the website format as a type of interactive gallery space for collaborative art and movement installations. One student explored the Zoom platform as a space for interactive audience engagement in solo performance, inviting audience members to comment on the work in progress and vote for creative options presented by the choreographer, which then shaped and informed the subsequent stage of the creative process.

As we tentatively emerged from some lockdown constraints, another student created and presented an immersive (socially distanced) installation work that combined drawing and dance. Significantly, the work contained a number of discarded objects, drawings, and materials that, whilst explored through process, never made it into the final performance composition but were nevertheless retained and incorporated into the work as material markers and traces of the research journey. This decision reflects the student's enhanced engagement with process (as opposed to product) as a key material central to the realisation and depth of this creative encounter, an attentiveness and awareness that arguably might not have emerged in "normal" circumstances and conventional making/production modalities. Alongside these specific examples, many excellent screendance works were created, providing students with a permanent record of their creative practice and performance skills. In addition to technical recording and production skills, learning skills of adaptability, creativity, problem-solving, and resilience evolved in new and unusual ways, skills that will hopefully stay with a generation of students for life.

Conclusion

By proposing and encouraging the act of making a dance that might never actually happen, we improvised with the constraints of linear time, educational schedules, and the directive towards an end goal or set date for completion or deadline in which the ongoing process of creation is executed, laid to rest, and ceases to exist. As opposed to engaging in a dash towards an end point, this process of making a dance that may never actually happen paradoxically produces a state of engaging with a dance that manifests, materialises, and *shows up* on a regular basis in the present through imaginative wandering, creative dreaming, and speculative fabulations—in a sense, it is a dance that *is always* happening.

As a result, the students' notebooks and process diaries were noticeably more creative, denser, and populated with multimodal forms of documenting experiences, such as drawing, diagrams, poetry, photography, capturing movement scores, evaluations of their outputs, and creative writing. Alongside the documentation, some in-depth reflection, analysis, and evaluation of process began to develop, as opposed to the more usual commentary on outcomes, products, and the end points of the creative and making process. The possibilities of imagining, creating, and speculating in these ways are therefore multiple and liberating, and, by association, unnerving, unsettling, and, in the contemporary neoliberal Higher Education system, challenging and often unwelcome.

As lockdown restrictions were lifted, however, assessment outcomes in the form of practical projects and performance outcomes were eventually delivered, and an inevitable return to a product-oriented mode of assessment ensued. However, for a brief time, we dwelt on the possibility of making and evaluating dance otherwise, through the articulation of speculative fabulation and creative dreaming, where anything was simultaneously possible and impossible, and creative imaginations took flight. Pedagogical recommendations emerging from this period are speculative at best; however, what remains for my own practice and might be useful to others is an appetite for creating space and time for exploration and experimentation. This sits alongside a desire to involve students further in dreaming, play, and speculative fabulation, and to value the materials at hand—taking time to acknowledge what shows up in the moment, to carefully nurture ideas and follow their trajectories, deviations, and detours, and to value and be led by the unknown, the unfixed, and the improbable.

References

Abrams, J., & Sack, D. (2020). Prologue—'On emergency'. *Imagined Theatres Journal* [online]. Retrieved June 1, 2024, from https://imaginedtheatres.com/emergency/

Amsler, S., & Facer, K. (2017). Contesting anticipatory regimes in education: Exploring alternative educational orientations to the future. *Futures, 94*, 6–14.

Bachelard, G. (1994). *The poetics of space.* Orion Press.

Bannerman, C., Sofaer, J., & Watt, J. (2006). *Navigating the unknown: The creative process in contemporary performing arts.* Middlesex University Press.

Claxton, G. (2006). Creative glide space. In C. Bannerman, J. Sofaer, & J. Watt (Eds.), Navigating the unknown: The creative process in contemporary performing arts. London: Middlesex University Press.

Egert, G. (2019). Everyday abstractions: Immediation and the powers of choreography in manning. In A. Munster, B. M. S. Thomsen, & E. Manning (Eds.), *Immediation II* (pp. 443–454). Open Humanities Press.

Gibson, J. J. (1979). The ecological approach to visual perception. Boston, MA: Houghton Mifflin.

Grosz, E. (2011). *Becoming undone: Darwinian reflections on life, politics and art.* Duke University Press.

Haraway, D. (2016). *Staying with the trouble: Making kin in the Chthulucene.* Duke University Press.

Hunter, V. (2021). *Site, dance and body: Movement, material and corporeal engagement.* Palgrave.

Le Guin, U. K. (1976). *The word for world is forest.* Orion Publishing.

McCormack, D. (2008). Thinking-spaces for research creation. *Inflexions, 1*(1), 1–8. http://www.inflexions.org/

Nelson, R. (2006). Practice-as-research and the problem of knowledge. *Performance Research, 11*(4), 105–116.

Nelson, R. (2022). *Practice-as-research in the arts (and beyond)* (2nd ed.). Palgrave.

Neri, L., & Respini, E. (2018). *William Forsythe: Choreographic objects.* Prestel Publishing.

Sheets-Johnston, M. (1969). *The phenomenology of dance.* University of Wisconsin Press.

Trimingham, M. (2002). A methodology for practice as research. *Studies in Theatre and Performance., 22*(1), 54.

Truman, S. E., & Springgay, S. (2015). The primacy of movement in research-creation: Newmaterialist approaches to art research and pedagogy. In M. Laverty & T. Lewis (Eds.), *Art's teachings, teaching's art: Philosophical, critical, and educational musings* (pp. 151–162). Springer publishing.

Truman, S. (2018). SF! Haraway's situated feminisms and speculative Fabulations in English class. *Studies in Philosophy and Education, 38*, 31–42.

van Heeswijk, J., Hlavajova, M., & Rakes, R. (2021). *Towards the not-yet: Art as public practice*. MIT Press.

Verger, N. B., Duymedjian, R., Wegener, C., & Glăveanu, V. P. (2024). Creative preservation: A framework of creativity in support of degrowth. *Review of General Psychology, 28*(3), 268–281.

White, C. A. (2002). Practice as research: Knowledge how and knowledge whether. *Contemporary Theatre Review, 12*(part 4), 113–120.

CHAPTER 8

Troubling the Recovering Academic: Poetry as Material Resistance

Portia Ungley

Introduction

> To tell the truth, we have to trap appearances with quotation marks. We are not what we are said to be—*Hélène Cixous*

This chapter is a call to arms—what kind of academic are you? How do you position yourself in the prevailing winds of contemporary academia? Do you feel safe and valued in the academy, ready to preserve and develop knowledge? Or are you an "academic", a member of the precariat, writing funding bids before research, productivity before knowledge creation? Are you in recovery from the addiction to the ivory tower, which has not existed for many decades? And how, in the name of ORCID, can we align these things? I offer my poetry as a way of making sense of and preserving the creative and pedagogical moment, and of opening it to change. It is also a call to accept the rigour of disruption, the trapping of our academic appearances in quotation marks. This is not creative as explanation or

P. Ungley (✉)
Faculty of Education, University of Cambridge, Cambridge, UK

N. B. Verger, D. Mattson (eds.), *Possibility Studies and Societal Futures*, Palgrave Advances in Possibility Studies,
https://doi.org/10.1007/978-3-032-26379-7_8

decoration; this is creative as onto-epistemological resistance (Barad, 2014; Brown et al., 2025; Verger et al., 2024).

To address this uncomfortable subject, I built a series of creative Kaprowian "happenings" to unveil the trouble of the academe; being present and researching in the institution of critique, and more particularly, how to render that space safe ***and*** brave (Arao & Clemens, 2013). Happenings are participatory, tightly organised but open-ended events which ask for engagement and bravery from their audiences. The happenings allowed for the occupation of the space of academia in the face of the economic turn, increased demands for impact and the ever-decreasing capacity for humanity. It also pointed towards ways of responding creatively. I'm so angry I made a sign, as ethnographer Micheal Taussig wrote in response to the Occupy Wall Street Protests (Taussig, 2012): this positioning of protest as holder of society's knowledge is dangerous, rebellious, an act of disruption to the status quo, the definition of the post in postqualitative (Aghasaleh & St Pierre, 2014). In this chapter it is also a way of preserving and enacting societal change as a mode of knowledge production, one which is certainly speculative but also radical in its anarchy (Gilman & Tipper, 2024; Manning, 2016).

The role of the personal and the affective is often viewed as less rigorous than that of its objective colleagues in research, but, drawing on the descriptive (artistic) versus the narrative (data-driven) school of autoethnography, I occupy this chapter as an education researcher and as a poet. I occupy this space to unravel the addiction of the academic—to research grants and impact statements—and offer interrelated joys, glimmers and anarchy as a path of recovery (Adams, 2012). Glimmers are used by the neurodiverse community as the antonym to triggers; in this broken academia, they are data that glows, shimmering moments that catch your breath, joyous conversations that change the world (MacLure, 2013; Rose, 2017).

They are also the resistance to materialising or reifying the academe into a neoliberal force for free market capitalism, another ground for training artificial intelligence on the labours for which we are not sufficiently or honestly remunerated, "generating coercion, exploitation, and submission through technocratic tools" (Verger et al., 2024, p. 275). It could be asked who is paying for this book's labour? And for the change that it seeks to enact? Does social justice represent electricity bills paid in its enaction? This is anarchic, as you, my fellow (re)searchers, are trained to stick to your discipline, stay in your lane and ask for anarchy ***within constraints*** (see Hunter, Chapt. 6, this volume). The idea of a beautiful constraint as "a limitation or defining parameter often the stimulus to find a better way of doing something" is an ambition and generative spark that I admire and use in my poetry, but there is something disturbing and uncomfortable in naming the boundaries to knowledge that confine us (Morgan & Barden, 2015, p. 11). Asserting the role of boundaries in this way stimulates questions around what constitutes education and its goals; if we give it a name, if we define its limits, the discomfort of what defines learning cannot be ignored. Are we serving the politics of education or reaching towards knowing ourselves? (Ergas, 2017).

Addressing these existential questions demands visions of a future; rewilded, reformed, reimagined (Cremin, 2025; Hay, 2025; Michael & Michael, 2025). Let's invite these ideas to percolate into and through each other—between them they embody the idea that "[we] think in words; for [us], language is an inexhaustible thread [we] weave as if life were created as [we] tell it" (Allende, 1999, p. 3). Join me at the future's loom to consider this as a form of recovery from the pressures of academic life, a way to start your healing, spinning a thread together.

This act of contiguity holds space for academics to explore their own recovery in four ways:

1. to upcycle the emotions and knowledge already created without destroying them;
2. to bricolage the methods of holding and preserving into footprint-light, impact-heavy revisions (Kincheloe, 2011; Levi-Strauss, 1966);
3. to fixpert, making said ideas accessible through the use of what low-fi resources and what knowledge already exists (Micklethwaite et al., 2020); and
4. to differentiate between the craftsmanship of risk and certainty (Pye, 1971).

These four forms of creative preservation are situated within the happenings, aligning with the creative preservation framework (Verger et al., 2024). The materials support the interrogation of the academic's resistance to the academe. They clarify the autoethnographic and postqualitative inquiry methods for the coinhabitants of the academe, drawing on transdisciplinary ideas and the tacit, discipline-specific training of the contributors—psychology for the authors of the creative presentation framework, the art school for the author. This kaleidoscope of ideas raises questions about the identity of those researching and calls for a rebellious form of research.

Equivoice as Academic Glimmer

Hélène Cixous drew on the energy of the May 1968 rebellion in the universities and streets of France, in which she participated, when she asks who am we? Postqualitative voices question the possibility of letting ourselves exist within the business of education. Can we, as Cixous' inheritors, misbehave beautifully, as I just did by questioning the ground of the book that these words exist in? (Cixous, 1976; McGregor, 2012). Dare we be the medusa that laughs when resisting? Cixous' energy of joyful refusal unravels the material of the university, questioning whether the warp of the institution and the weft of the research are creating a higher education of which we should be proud and more so demands that we question our position through the interrogation of

> … what touches you, the equivoice that affects you, fills your breast with an urge to come to language and launches your force; the rhythm that laughs you; the intimate recipient who makes all metaphors possible and desirable (Cixous, 1976, p. 882).

Talking of the atmosphere of a historical textual moment, and already two rebellious moments, is an exercise in honesty of the me that I bring to this chapter, to the happenings and to the poetry I risk in this space of social science. The language of the artist gives traction to Cixous' interrogative equivoice, launching my force, and holding the possibility to make all metaphors. Metaphors are "general map pings across conceptual domains" and poetry is ripe with both metaphors and conceptual spaces; of reflection on modes of knowing intra-fears, intra-laps and intra-acts—it is explicitly personal as we engage with our onto-epistemological groundings (Lakoff, 1993, p. 202).

If poetry is to be accepted as a mode of knowledge creation and preservation, it is important to refine the role of narrative, differentiating between meaning-giving and experience; however, this is easily undermined if the goal of certainty is prioritised. The mode of knowledge can be explicitly (rather than tacitly) experiential and Cixous's equivoice can be privileged as a desired outcome. Histories can be written within narrative structures "storytelling on different levels … takes part in weaving, unravelling, and reconstructing the fragile fabric of our narrative existence" (Meretoja, 2014, p. 105). This is an unravelling, a deliberate engagement with the uncomfortable. These radical methodological breaks exhort us to read through theory, and challenge the conflation of rigour with rigid, serious with sophisticated, embedded with useful; it is an exercise in "imagining new becomings, returns, new departures, and detours" (Denzin & Lincoln, 2018).

Poetry, Risk and Art

Poetry is the material of this paper. I offer four poems: three developed with participants including professors and early career researchers, artists and engineers, cynics and believers of the importance of risk. They were gleaned in the *in-between* spaces of lunchtimes and early evenings (Gregg & Seigworth, 2010). What became clear is that the role of risk and imagination projects risk working as an impetus, while the positionality of the soul responds to that which is uncomfortable. The drive for safety can lead to a suffocation of autonomy (Dufourmantelle, 2019, p. 33). This raises the idea of use versus construction, and what it is to be "of" the construction. When collaborating as a fellow coinhabitant of the academe, but also the impetus behind the creation of a happening.

This is a critique of those who use arts as decoration—while there is value in art therapy and other playful incursions into the creative milieu, there is a difference between art and the decorative. This is explored in Thomas Merton's meditation on the art of shipbuilding in which the idea that you can go to sea in a ship, in admiration of the vessel and its maker, but without understanding its construction—"go to sea without art" (Merton, 2009, p. 10). Yet, there is a risk of slipping into the positioning of the ship itself instead of acknowledging and praising the skill with which going to sea is achieved—"nothing but approval for ships and shipbuilders, ... no patience with idols and their makers". As such, in this chapter, the poems are my ships and shipbuilders, but my methodology is a rejection of reifying them as either art as decoration (the idol) or an arts-based research project (the makers). The art of going to sea, in the context of this chapter, is using and accepting poetry as a thing in its own right, and the collaborators in its construction as having part ownership of that thing. The acceptance of poetry as a material of resistance against the *status quo* is the key methodological standpoint.

However, this standpoint is developed from a point of theoretical rigour. This combination of going to sea with art but also being aware of the skill and techniques (rigour) that creates said art (poetry and autoethnographic postqualitative inquiry) offers a way for the academic to be in recovery. It is born from the particularly affective enactment of Dufourmantelle's praise of risk, in which she lost her life saving strangers from drowning. This chapter is not zero risk; it has gone to sea *with* art. The oscillation between risk and wisdom speaks to the addictiveness of such risky behaviour, even when that behaviour could be classed as spiritually grounded, altruistic or innovative/original. Between them, Dufourmantelle and Merton ground the affectiveness of the methodologies which underpin this chapter.

This affectiveness embraces the possibility of thick data, "the detailed account of field experiences in which the researcher makes explicit the patterns of cultural and social relationships and puts them in context" (Geertz, 1973, p. 6). The materiality of the experience itself is part of data collection, but often relegated to memos or journals, secondary to the data for analysis. The happenings that I constructed were tightly planned and highly participative to the extent that I gave up my leader/ maker role in pursuit of the participants' ideas by ceding control. As Allan Kaprow discussed when creating his happenings, "all of life will be open to [the young artist] … out of nothing he will devise the extraordinary and then maybe nothingness as well" (Kaprow, 1958). In this context, the participants are the young artists, my role is to be the autoethnographer who actively listens to their ideas, intentionally holds space for the issues raised and gently sits with their trouble (Cave & Sloan, 2014; Haraway, 2016; Winnicott, 1960).

The poems were created based on tropes of the academic—addicted, recovering, teen, pilgrim, warrior, joyful—each addressing particular aspects of the academic life. Each happening was based on an artistic intervention, deliberately low-fi; this one was a playful engagement with paint chips of the sort you find in DIY stores. Choosing names from the hundreds of options, each coparticipant began to make sense of the words chosen, morphing them into a narrative, structuring their knowledge affectively. This bricolage of words and ideas was worked into the beginnings of a poem—a stub, a fragment. Some were more focused than others—the most colours chosen were 21, the fewest 6. I took these poetry seedlings and developed them into full poems, not to analyse, but as a way of transferring the space that was held during the interaction, the sitting with of the trouble of contemporary academia. As the collection of poems expanded, I drew on Alexander Calder's mobiles, creating moving sculptures from the paint chips, delicately balanced. Suspended from discarded tripods and parasol skeletons, the mundane paint chips became part of *poetrees*, which I exhibited, the poems in the catalogue and taped to the floor beneath the spoked branches. The humble paint chips were free, the armature second hand (see Picture 8.1).

The low cost and accessibility of materials are echoed in the accessibility of creating found poetry—not having to choose the words removes much of the reluctance of the academic who does not write poetry to engage

Picture 8.1 Photograph of poetree taken in October 2024 in the Great Hall, Homerton College (credit Fabiana Zonca and Carlo Nicora 2025)

with the process. The texts created are therefore both accessible in their creation but also engaged in their honesties. The ideas were re-lensed through compassionate autoethnography, capturing emotions and filtering them into moments expropriated as things. Many of those moments were challenging—failed collaborations, anxiety for success, challenged identities—but "just when you think you can't stand the pain anymore, well that's when the real work has only begun. Then there's the vulnerability of revealing yourself, not being able to take back what you've written or having any control over how readers interpret it" (Ellis, 1999, pp. 671–672).

In this chapter, I offer three poems as examples of this anarchic, affective and vulnerable knowledge. The first poem I offer was a pilot by me, as I first engaged with the process, full of nerves and uncertainty.

emotion wheel
shelled drifts of nettle veils
borrowed richness, mouse-fragile
fresh thunder buried fossil deep
crunching electric salt and dust

paint chips names: shell stories, denim drift, sea nettle, delicate veil, borrowed blues, rich black, mouse tail, fragile blossom, fresh powder, thunder cloud, buried treasure, deep fossil, caramel crunch, electric storm, rock salt, chalk dust

This short poem allows me to relive the emotions of that first happening each time I read it. It is also the poem that was the most prescriptively kept to the use of words from the colours—the only two words of nineteen that are not in the colour names are "of" and "and". It was a useful exercise in seeing that too strict rules, however much constraint is beautiful, can lead to unhelpful limitations, and I relaxed the rules I set for my fellow co-inhabitants as a result. The imagery is multi-textural—there is taste and texture and pain as well as anthropomorphising of feelings. This example is easiest to describe as it is an autoethnography of my own emotions—and it serves as a reminder of what it is to take risks.

The first line describes the feeling that I was hiding and that the act of hiding was painful in the stings of nettled veils. The second holds the tension of possibility with the fragility of uncertainty. The third returns to the need to conceal, the fresh thunder of my heartbeat masked as I powdered my own reflections into poetry. The final line draws on the way that the sweet treats I brought to welcome my participants were tainted with nervousness and disquiet. These feelings are perhaps not obvious to the reader, but as a piece of affective witnessing, I went to sea with the art of my process, while writing through the discomfort of exposing that part of me.

In the next, I was challenged by the idea of ethical affective expropriation of ideas from a coinhabitant of my own academic institution. When considering purposive or snowball sampling, the idea that the participant is chosen due to knowledge of their expertise is justified amongst ethics and consent forms. In this case, there is no such framework. As the colours were chosen and the conversation unravelled in different directions, my personal knowledge and friendship led me to question how, in the generation of insider knowledge, "the addressee may have to temper her or his response in order to accommodate my presence" (Adams, 2012, p. 186). Was it possible for my own personal insight colouring the creation of the poem?

mother's chalk
annotating in breast milk
inscribing lines to tailor
the next generation's leaps
alarm clock shrilling
a call to different arms
needling for change
flowering with blurred hope
possibilities in misty mirror
from home's palm nights
to maybe's nordic skies
willows weeping for the forgotten
for education's fallen

paint chips names: tailors chalk, primrose white, leap frog, natural wicker, pine needle, passion flower, misty mirror, palm night, nordic sky, willow tree

This poem talks of the joys and difficulties of multiple identities within the academic framework, specifically polyglot playfulness around language (wicker transposed to the German of Wecker and retranslated as alarm clock) and conversations of Cixous' white ink of breastmilk in relation to tailor's chalk. The richness of free association, of unbound discussion, was limited *less* by friendship, and although Ken Gale allows himself to be "trembling with matter", I find myself drawn to the old paradigm wars of objectivity and subjectivity, and the need to delineate between myself and my participant, even as we play-together-apart, write-together-apart, suffer-together-apart (Barad, 2014; Gale, 2018).

The shared experiences shine through the words, of being mother and revolutionary, of accepting role of outsider and neurodiverse. There is so much knowledge caught in these lines of poem, found in the names of paint I will never use. There is compassion and relief in this poetry, and subversion of the rules and norms that I developed and relaxed. The metaphors are too densely packed to just ping across conceptual domains; they are dancing and leaping; they are anarchically joyous. The use of movement as a way of exploring emotion is established, but the idea of losing oneself in the process draws on a particular freedom associated with the construction of bodily knowledge.

> I danced till I disappeared inside the dance, till there was nothing left of me but the rhythm of my breath. And in the rhythm of my breath I felt totally connected body and soul. (Roth, 1997, p. xxiii).

The idea that the autoethnographer can lose herself in the writing of the experience, can be willingly and perhaps wilfully subsumed into the experience, is core to the postqualitative research definition, that "rather than observing and documenting a river and its contents from a fixed point on the bank, [it] takes to a boat and becomes part of the flow it wants to fully understand" (Fox & Alldred, 2022, p. 10).

This moment of gleeful diving into the flow of the research topic defines my version of postqualitative descriptive autoethnography, and while the first poem is the hauntology of my academic self by past engagements, the second is the "virtuous upward spiral" of joy and gratitude, experience shared (Johnson, 2020, p. 7).

The final anarchic mode of knowing came through a stolen moment, a compliment on a garment and a need to capture the joy of that encounter, by both of us. As a creative writer, this academic had skin in the game, a desire for the possibilities offered by such a speculative fabulation to be captured, quickly, without too much overthinking.

drama flowers
disordered cowslips, contained by geometry
passing unseen by the flow of knowledge
violet petals pressing to the infinite
comfortable riverbed, water smoothing
sacrificing the eternal for the fresh
flint arrowhead piercing bucolic bliss
under the wheels of Lady Penelope
sparks flaring as stones metal-skid
dewy lawn rent with muddy destruction
skinning knowledge for jewelled kisses
of mutable, reflexive possibilities

paint chips names: cowslip flower, violet kiss, pressed petal, river bed, flint arrow, Lady Penelope, dewy lawn, violet jewel

The poem is less reflective in some ways, less coloured by the question of addiction and recovery in the academe as by the warmth of the interaction, two brains enjoying the excuse to play with words. The choice of one colour name set the tone, Lady Penelope, her pink Rolls Royce with machine guns spraying stones as the sharp and punk counterpoint to otherwise natural imagery. The texture of the words of the poem holds the grin of recognition—a fellow rebel hiding in plain sight (Burnard et al., 2022). The juxtaposition of this hidden knowledge in the gentleness of the names chosen highlights the effectiveness of this method of eliciting affective encounter. There is no explicit anarchy in the choices yet the poem's mind's eye is full of woad-covered warriors hiding in the bend of the river, of the latent feminism of Lady Penelope's stilted puppet movements as she took care of what the heroes of the Thunderbirds had not, of the abandonment into colour when the battle is won. The memory of a conversation touching on possibility, creativity and defiance lives in this poem and is therefore transmissible in its capture. How else could such an encounter be captured?

Poetry, in its ability to capture ephemerality, allows a resistance to exactly what I am doing—writing a moment into a textual record devoid of its context. You have met an affective versioning of the ideas of the unravelled academic in this chapter through the choice of epistemology, but also in the moments recorded in poetry—Isabel Allende is my favourite fiction author, and because of her "language is an inexhaustible thread [I] weave as if life were created as [I] tell it"—her words changed me and influence the poet I am today but also the author you encounter in this chapter. As Elizabeth Mackinlay used epistemic friendships in letters to Virginia Woolf and Simone de Beauvoir to explore her feminist autoethnographies, these poems are palimpsests of encounter, transcription and memory (Mackinlay, 2022). Combining atmosphere with storytelling and unravellings demands the use of joyful autoethnography, underpinning a heretical, personal and joyful dive into what academic research methods can look like if we stop gatekeeping, start embracing diversity and honouring different forms of knowledge (Adams, 2012).

Trapping Knowledge in Disruption

The question that this raises is what this disruption means to data, particularly as we move to an era where the more-than-human is mistaken for the less-than-human of AI. As a postqualitative and feminist autoethnographer, I am often quizzed as to where the limits of my field lie, if anything goes or whether this is how I see the future of academic study. A simple answer is in Aghasaleh and St Pierre's (2014) definition of postqualitative inquiry, that it is "impossible to bracket out prior knowledge, experiences, or feelings ... the researcher does not exist prior to doing the research ... it's purpose is rather to destabilize and question the normativity of conventional research methodology in social sciences".

Walking my talk, sitting with the trouble of this distillation of her position, this chapter champions the need for the inclusion of previous explicitly human experiences, as evidenced in Allende's inexhaustible thread. In the performance of the happenings, I disappeared inside the role as coinhabitant, born as this version of myself through their enaction. However, would I ask for all research to be undertaken this way? Is the disruption of the academic in any way my goal for future academic study? Is it speculative pragmatism? Is it "interested in the anarchy at the heart of all process, and is engaged with the techniques that tune the anarchical into new modes of knowledge ... must [it] emerge from within the occasion of experience, and cast out from within its formation the stakes of its coming-to-be" (Manning, 2016)?

The coming-to-be echoes "a space of interrelated intelligibility articulations that shape these particular possibilities in advance", the materiality of Heidegger's *being-in-the-world* and *worlding* (Keane, 2020, p. 282). The interrelationality of forming and being formed by the context rapidly transforms into the intrapersonal engagements of speculation and pragmatism—they are not separate to each other but intimately entangled. The textual component of the postqualitative inquiry is not separate to the subject researched, it is created in its friction with its creation, ouroboros, chicken and egg. However the interrogation of modes of knowledge is "not '*I* am wondering' but 'Where does this movement wonder me?" (Manning, 2014, p. 167). In the act of research, the act of artistic intervention, the intra-action, the answer to the question as yet unasked, it found—what is this chapter exploring and why is it important?

The textual transference of being in the world of academe is mitigated against the creation of it—this form of research, creation, practice allows for the recovering academic to be both wondering in the fields of disciplinary specificity and to be shaped by said movement. The space for such speculation asks "what techniques of existence are called forth by our singular ways of engaging and being engaged by the world" and, more so, "how can we create an environment where knowing as final act is not the aim? Where what is remembered is *how* we came to thought" (Manning & Bozalek, 2024, pp. 2, 4 emphasis in original).

Knowledge creation and the trapping of it is therefore necessitated beyond the answering of research questions—what is this chapter doing? Using poetry to capture affective moments around a core concern of the academe. Why is it important? Because the current practices do not allow the space for the thickness of experience to be held in comfortable tension with the academe itself. Speculative pragmatism is a deliberate act of anarchy in response to the need to pragmatise and positivise in order to consider originality a valid and rigorous contribution to knowledge.

The risk of this quantification of knowledge and its value can be understood as strawman in which academics dream of SCOPUS scores and journal rankings, in the model of Philip K. Dick's electric sheep (Dick, 1968). Is this method of understanding the value of research subsuming the humanity of the academic, until the statistical importance of data and its publication is more valuable than its purpose or contribution?

From this, knowledge and its mode have an identity, an ontoepistemolgical private life, and are redolent of Cixous' personal questioning. She asks "what is a human subject, what is it that makes us live so well and so badly … I never ask myself 'who am I?' (qui suis-je?) I ask myself 'who are I?' (qui sont-je?)" (Cixous, 2003, p. xvii). The Is that I have brought to this chapter are bolstered by the theorists who allow epistemological anarchy in order to offer speculative space, to sit with the trouble of questioning the notion of knowledge, of unravelling the recovering academic. The act of unravelling is not an act of destruction as misunderstood in the post of postqualitative, but a disruption; returning to the three keynotes at Possibility Studies Conference held in Cambridge, 2024, a rewilding, a reforming, a recycling. The new mode of knowledge offered is softer, more human, more than human, or as Nora Bateson describes it.

> We exude warm data in our eyes, our smiles, our verbal and non-verbal conversations and in the importance we accord to being in relationships with others at the personal, family, community levels and in local, national and global contexts. We also have the capacity to tap into ancient wisdom to add to our store of warm data and the capabilities that get unleashed when I see you in me and me in you (Bateson, 2023, p. 24).

I imagine my poetry to be this mode of knowing, a warm, glowing, thick, shimmering affective glimmer. It is an invitation, an invocation, a call to arms. Will you accept or reject it as a way, one of many, of bringing possibility back into the academe and preserving it? It is not the only future but one in a space in desperate need of options. Over the last few years, the world has seen the closure of departments, funding cuts, organisational restructures on a scale beyond imagination, quite apart from the resurrection of Nixon's idea that "universities do not pursue knowledge and truth, they pursue deceit and lies" (Vance, 2021). The academe is under threat, but which bits do we want to save? Do we dare admit there are parts that could crash into the sea, such as the cliffs of colonialist, classist and ableist mores? How do we move forward from/with the economic turn? How do we protect from the bogeyman of AI, not only for our students but our own research? What can stop the thread of the unravelling academe from being pulled and leaving academia with nothing but many billion characters of word salad?

Enaction and Possibilities Meditation

These questions are necessarily rhetorical, and this chapter does not dare to offer a definitive answer. However, it does offer a synthesis of identity building and knowledge-making through autoethnography. It holds the space for me to inventory the traces of the human, to embrace the material collection that is the heart of our profession and to formulate a resistance to the narrowing possibilities, a counterpoint to the neoliberal desire for discreet, concrete and boundaried answers that weeds out serendipity and ignores glimmers. The data that is sometimes ignored—the warm, the personal—is the unwanted growth of dandelions and buttercups, the weeds in the garden of academia. Pain and grief are welcomed into this garden—the hauntologies of stories of loss, shared in stolen moments within the academic rhythm, not catalogued by name, but by emotion. Post-pandemic, the unravelling of the academic is more visible, our human fragility impossible to ignore, its acknowledgement a consideration on the state of the academe.

The final poem is the pilgrim's meditation written from the glimmers offered by conference participants as part of a meditative yoga practice. Poetry offers a window into postqualitative research gathering—it is not poetic inquiry, but the use of the butterfly net of theory to capture affective shimmers. The energy of the space, the participants, the desire to break new ground showed that interdisciplinary, non-hierarchical, affective and friendly environments are a workable alternative to computer banks and silence. This chapter is not going to gloss over how important the people at the Possibility Studies Conference 2024 were to its development—not as a mass of heterogeneous academics, but as individuals with lives and stories, encompassing Adams' pain, confusion, anger and uncertainty (Adams, 2012). They are the basis of an axiology of affective expropriation, developing ethical and compassionate autoethnographies in emotion-driven research (Edwards, 2021; Ellis, 1999, 2016; Ungley, in press).

To draw out the emotions, I invited coinhabitants of the academe to join me after a long day of possibilities, bringing a text they loved for gentle stretching. The impetus is movement-based elicitation and Maclure's data that glows in favourite texts—one of the risks and joys of such an intervention is the lack of control of the outcome (Dietrich,

2017; Ditzel Facci, 2020; MacLure, 2013). Across an hour, the laughter and intimacy grew. The pilgrimage is the movement towards something in which the process becomes more important than the destination. It is a metaphor for the shift towards postqualitative inquiry through autoethnography, the conceptual pings are a match. The texts that were offered ranged from Mary Oliver—anything by Mary Oliver, to the House on Pooh Corner via It's a Wonderful Life and Forest Gump. The glimmers that were brought were entirely personal, but no one brought an academic text. It felt like a bleak moment to realise that the metier in which the coinhabitants of the academe live and the outputs that they create did not constitute the texts that brought joy. As a poet, I use words as the joy-buffers for the harshness of the academe, and that the coinhabitant choices reflected the need to wrap ourselves in words that bring glimmers, not triggers. Is it unrealistic to expect wordsmiths to like their own words?

The academic pilgrim responds to this mismatch as an exercise in kindness. The need for this is invoked in postqualitative and feminist tests: "Rather than finding despair, paralysis, nihilism, apoliticism, irresponsibility, or immorality in the decay and devastation of the ruins ... [we] have found possibilities for different worlds that might, perhaps, not be so cruel to so many people" (St. Pierre & Pillow, 2000, p. 1). The process of kindness begins in the etymology of pilgrimage, the peregrination of someone from another field, of wayfarers, of travellers, of crusaders. The unfamiliar is welcomed in the act of pilgrimage and the process-based journey, core to the *disconcertions* that allow space for instinctive responses and their affective shimmers (MacLure, 2011, 2013; Taussig, 2018). The ruins of the ivory tower were mentioned at the beginning of the chapter but remerge in St. Pierre and Pillow's discussion of poststructuralism and in Maclure's questioning of the perceived destruction of qualitative research. This poem is an attempt to imagine the shift from recycling (making different) to upcycling (building from) of what is no longer useful.

The pilgrim's meditation

our feet calloused by motion	our hands reach out to feel leaves	voices raised in song, a chorus of affirmation
voluntarily hobbling our skins	new experiences caressing bare skin	possibilities rather than problems
blisters and bunions and corns	catching fragrances unknown	discernment over dissonance
dirt embedded in poorly kept nails	in gentle exploration of texture and experience	a thousand thousand yes-and's
so is our daily grind for perfection	so is our flourishing in the unknown	so is our shared uplift of soul
our clothes have stained hems	breaths joined in discussion canons	glimmering, glowing moments of hope
we smell of the road we travel	ties formed across canyons of doubt	down rabbit holes of positivity
of dusty paths we have woven	wrists clasped, eyes locked, promises offered	acceptance of the vagaries of life
of lush fronds trodden down	forgiveness in awareness, open handed	maybe kingfisher-darts, gaining weight
so is our search for pilgrimage	so is our raising of structures	so is the deep draught of our mayhap
our throats are dried by passing fields	shared water, simple imagination gestures	kneeling and kissing the ground of our daring
our stomachs hungry for sustenance	eyes clearing from the veil of uncertainty	leaning far into intuitive movements
to feed our needs, for easy digestion	apathy abandoned for engagement	dancing each other's iron cages, flames flickering
but in faring together, much sits heavy	collective joy brimming empty cups	smudging history's status quo in wet time
so is our reluctant futures sharing	so is our forgiveness of our selves	so is our beautiful misbehaviour in glad grace

The use of the term meditation in the title of the poem is a nod to the religious context associated with the term pilgrimage and also to the use of yoga pranayama (breathwork) as the means of drawing out ideas, influenced by transrational peace studies for conflict transformation such as those using dance as an elicitative method (Dietrich, 2017; Ditzel Facci, 2020). Hilary Cremin's aesthetic considerations of peace building, particularly "the role of higher education peace scholars in perpetuating violence" culturally, structurally and poststructurally, allow for this chapter's consideration of how the academy has perpetrated its multivoiced violences on the academic body and questions the journey away from structurally unsound solutions (Cremin, 2016; Kester & Cremin, 2017, p. 1416).

The nine stanzas of the poem are footsteps on a path, the journeying from dis-ease to grace, through process. In the first stanza, the suffering which is intrinsic to the process of discovery is addressed and its associated ill, lack of self care. In the second the causes of stains and tears are laid bare, the use of hopes and dreams to line the road. It brings to mind Yeats' poem *He Wishes for the Cloths of Heaven* in which he reminds us we tread on his dreams, because the academic's recovery is predicated on the need to acknowledge what has been sacrificed—idealism, youth, peace of mind. The third stanza is a recognition of the physicality of the need to contribute to original knowledge and perhaps a trace of what needs to be done. In the fourth, the pursuit of the unknown stops being a thing to fear, with a move towards experiencing the transitions from observer to participant. In the fifth, the acknowledgement of the presence and impact of those around the individual academic in the construction of knowledge draws on the image of raising a barn. The role of others is added to in the sixth and seventh stanzas, with self-acceptance intertwining with the role of possibility in collaboration. In the eighth stanza, the maybe of serendipity, the flashes of the unexpected, build towards satiety. The final stanza draws on Rumi's exhortation that there are hundreds of ways to kneel and kiss the ground, that the constraint of research is an arena to blossom within and that in challenging the accepted, there is space to sit with the trouble of our shared academe.

Each stanza is coloured by texts offered during the happening, through conversations and the intrinsic sense of movement. These modes of knowledge are explicit but also caught in the metaphor of pilgrimage. The elicitation of glimmers from coinhabitants of the academe under research is magnified in the constraint of poetry and the freedom of creative writing (see Picture 8.2).

When I exhibited this, I invited participants to lay on yoga mats, using bolsters and other props to get comfortable, under the neon sign "misbehave beautifully". Both an audio recording and printed copies of the meditation allowed immersion in the journey, under the vaulted ceiling of the Great Hall, Homerton College, Cambridge. It was a moment's stillness within the academic context but also external to it; a resistance in the

Picture 8.2 Photograph of the space to listen and engage with the pilgrim's meditation, taken in October 2024 in the Great Hall, Homerton College (credit Fabiana Zonca and Carlo Nicora 2025)

fabric of the academe but also part of its tapestry of possibility. The unravelling of negative emotions into calm allowed for questioning without conflict, elicitation of views without the need for agreement.

This enactment offers a moment of calm to the themes of disruption, resistance and anarchy because who we are is also reflective, engaged with ontoepistemologies which explore beyond the confines of disciplines and are enmeshed in ecologies which are both authentic and playful (Ingold, 2000; Sturrock & Else, 2007). We lose our jagged, spiky selves in the rhythm of the warm, glowing, glimmering affect of the human.

Poetry as Material Resistance

This chapter has woven theoretical references through the space held for poetic interventions, enabling them to be viewed as a mode of knowing. In sitting with the trouble of these new ways of knowing, it seeks to disrupt traditional approaches and offers an alternative to the materiality of data. In considering affect as equally rigorous as theory, this onto-epistemology has opened the possibility to a joyful, anarchic acceptance of the role of data beyond analysis. However, we will not go to sea without art, even in the pursuit of discovering non-traditional, disruptive but fundamentally rigorous modes of knowledge. Poetry as art, in the sea of academia, is a ship sailed by playful coinhabitants. Art is the materiality of its resistance.

This notion of disruptive rigour sits within the framework of creative preservation, holding space and sitting with the trouble of difference. It seeks to be AI and disenchantment proof, as it relies on the exceptionally human desire for agitation—anarchy resurfaces at moments of speculation and storytelling where knowledge is held and transformed, the academic wildside (Sinner et al., 2018). The ethical dilemma of what education is for is addressed in the acknowledgement of the importance of humanity and creativity. Serving the politics of education is secondary to the pursuit of self-awareness.

The academy is in need of upcycling but rigour, not rigidity, is key. The materiality of poetry (and other creativities) offers resistance to modes of knowledge which ignore the humanity that underpins the academic. The recovering academic is wounded but finds solace in the playful, joyful and its own affect. The warp of disruption and reflection weave through complex, interrelated and interdisciplinary concerns about the academic, held in place by the sinuous complexity of poetry's weft, containing the ability to expand and contract, safely and bravely. By replacing the materiality of data with the flexibility of poetry and art, this chapter offers a framework to support and enable the recovery of the academic and a route to its survival.

References

Adams, T. (2012). The joys of autoethnography: Possibilities for communication research. *Qualitative Communication Research, 1*(2), 181–194. https://doi.org/10.1525/qcr.2012.1.2.181

Aghasaleh, R., & St Pierre, E. A. (2014). A reader's guide to a post-qualitative inquiry proposal. *Google Docs.* https://docs.google.com/document/d/1oNTt35mQmDqiRaJot0rDbFdeaVOrlTcKK_9pV563LIY/mobilebasic?usp=embed_facebook

Allende, I. (1999). *The stories of Eva Luna*. Simon and Schuster.

Arao, B., & Clemens, K. (2013). From safe spaces to brave spaces: A new way to frame dialogue around diversity and social justice. In L. M. Landreman (Ed.), *The art of effective facilitation: Reflections from social justice educators* (pp. 135–150). Styles Publishing.

Barad, K. (2014). Diffracting diffraction: Cutting together-apart. *Parallax, 20*(3), 168–187. https://doi.org/10.1080/13534645.2014.927623

Bateson, N. (2023). *Combining* (1st ed.). Triarchy Press.

Brown, N., Ergül, H., & Power, N. (2025). *Social fiction: Bridging academic enquiry and creative writing*. Bristol University Press. https://doi.org/10.1332/30502969Y2025D000000004

Burnard, P., Mackinlay, E., Rousell, D., & Dragovic, T. (2022). *Doing rebellious research: In and beyond the Academy*. Brill. https://brill.com/display/title/62341

Cave, M., & Sloan, S. (2014). *Listening on the edge: Oral history in the aftermath of crisis*. Oxford University Press. https://www.semanticscholar.org/paper/Listening-on-the-edge-%3A-oral-history-in-the-of-Cave-Sloan/591d81bd4d5b368516035a78c370e0b4c0724cc4

Cixous, H. (1976). The laugh of the medusa (K. Cohen & P. Cohen, trans.). *Signs, 1*(4), 875–893.

Cixous, H. (2003). *The Hélène Cixous reader* (S. Sellers, Ed.). Routledge.

Cremin, H. (2016). Peace education research in the twenty-first century: Three concepts facing crisis or opportunity? *Journal of Peace Education, 13*(1), 1–17. https://doi.org/10.1080/17400201.2015.1069736

Cremin, H. (2025). *Rewilding education: Rethinking the place of schools now and in the future* (1st ed.). Taylor & Francis Group.

Denzin, N. K., & Lincoln, Y. S. (Eds.). (2018). *The SAGE handbook of qualitative research* (5th ed.). SAGE.

Dick, P. K. (1968) *Do androids dream of electric sheep?* Doubleday.

Dietrich, W. (2017). *Elicitive conflict mapping/Wolfgang Dietrich*. Palgrave Macmillan.

Ditzel Facci, P. (2020). *Dancing conflicts, unfolding peaces: Movement as method to elicit conflict transformation*. Palgrave Macmillan.

Dufourmantelle, A. (2019). *In praise of risk*. Fordham University Press.

Edwards, J. (2021). Ethical autoethnography: Is it possible? *International Journal of Qualitative Methods, 20*, 1609406921995306. https://doi.org/10.1177/1609406921995306

Ellis, C. (1999). Heartful autoethnography. *Qualitative Health Research, 9*(5), 669–683. https://doi.org/10.1177/104973299129122153

Ellis, C. (2016). Compassionate research: Interviewing and storytelling from a relational ethics of care. In *The Routledge international handbook on narrative and life history*. Routledge. https://doi.org/10.13140/RG.2.1.3129.0084

Ergas, O. (2017). Reclaiming ethics through "self": A conceptual model of teaching practice. *Teaching and Teacher Education, 68*, 252–261. https://doi.org/10.1016/j.tate.2017.09.013

Fox, N., & Alldred, P. (2022). *New materialism*. SAGE Publications Ltd. https://doi.org/10.4135/9781526421036768465

Gale, K. (2018). *Madness as methodology: Bringing concepts to life in contemporary Theorising and inquiry*. Routledge. https://www.routledge.com/Madness-as-Methodology-Bringing-Concepts-to-Life-in-Contemporary-Theorising/Gale/p/book/9781138066021

Geertz, C. (1973). Thick description: Towards an interpretive theory of culture. In *The interpretation of cultures*. Basic Books. https://philpapers.org/archive/GEETTD.pdf

Gilman, L., & Tipper, B. (2024). *Fiction and research: A guide to connecting stories and inquiry*. Policy Press. https://doi.org/10.56687/9781447369752

Gregg, M., & Seigworth, G. J. (2010). *The affect theory reader*. Duke University Press.

Haraway, D. (2016). *Staying with the trouble: Making kin in the Chthulucene*. Duke University Press.

Hay, P. (2025). The power of imagination. *International Journal of Art & Design Education, 44*(2), 308–312. https://doi.org/10.1111/jade.12585

Ingold, T. (2000). *The perception of the environment: Essays on livelihood, dwelling and skill*. Routledge.

Johnson, M. K. (2020). Joy: A review of the literature and suggestions for future directions. *The Journal of Positive Psychology, 15*(1), 5–24. https://doi.org/10.1080/17439760.2019.1685581

Kaprow, A. (1958). The legacy of Jackson Pollock. ArtNews https://www.artnews.com/art-news/retrospective/archives-allan-kaprow-legacy-jackson-pollock-1958-9768/

Keane, N. (2020). Dasein and world: Heidegger's reconceiving of the transcendental after Husserl. *Journal of Transcendental Philosophy, 1*(3), 265–287. https://doi.org/10.1515/jtph-2020-0012

Kester, K., & Cremin, H. (2017). Peace education and peace education research: Toward a concept of poststructural violence and second-order reflexivity. *Educational Philosophy and Theory, 49*(14), 1415–1427. https://doi.org/10.1080/00131857.2017.1313715

Kincheloe, J. L. (2011). Describing the bricolage. In *Key works in critical pedagogy* (pp. 177–189). SensePublishers. https://doi.org/10.1007/978-94-6091-397-6_15

Lakoff, G. (1993). The contemporary theory of metaphor. In *Metaphor and thought* (2nd ed., pp. 202–251). Cambridge University Press. https://doi.org/10.1017/cbo9781139173865.013

Levi-Strauss, C. (1966). *The savage mind*. University of Chicago Press.

Mackinlay, E. (2022). *Writing feminist autoethnography: In love with theory, words, and the language of women writers*. Routledge. https://www.routledge.com/Writing-Feminist-Autoethnography-In-Love-With-Theory-Words-and-the-Language/Mackinlay/p/book/9780367479770

MacLure, M. (2011). Qualitative inquiry: Where are the ruins? *Qualitative Inquiry, 17*(10), 997–1005. https://doi.org/10.1177/1077800411423198

MacLure, M. (2013). The wonder of data. *Cultural Studies↔ Critical Methodologies, 13*(4), 228–232. https://doi.org/10.1177/1532708613487863

Manning, E. (2014). Wondering the world directly—Or, how movement outruns the subject. *Body & Society, 20*(3–4), 162–188. https://doi.org/10.1177/1357034X14546357

Manning, E. (2016). *The minor gesture*. Duke Press. https://doi.org/10.1215/9780822374411

Manning, E., & Bozalek, V. (2024). In conversation with Erin Manning: A refusal of Neurotypicality through attunements to learning otherwise. *Qualitative Inquiry*, 1–17. https://doi.org/10.1177/10778004241254397

McGregor, W. (2012). A choreographer's creative process in real time. [TED talk] https://www.ted.com/talks/wayne_mcgregor_a_choreographer_s_creative_process_in_real_time

Meretoja, H. (2014). Narrative and human existence: Ontology, epistemology, and ethics. *New Literary History, 45*(1), 89–109. https://doi.org/10.1353/nlh.2014.0001

Merton, T. (2009). *Conjectures of a guilty bystander*. Crown Publishing Group.

Michael, M. G., & Michael, K. (2025). Editorial Axis of access—The controllers and the controlled. *IEEE Transactions on Technology and Society, 6*(3), 232–241. https://doi.org/10.1109/TTS.2025.3567737

Micklethwaite, P., Charny, D., Alvarado, M., Cassim, J., Dong, Y., & de Vere, I. (2020). *Fixperts: Models, learning and social contexts.* DRS Biennial Conference Series. https://dl.designresearchsociety.org/drs-conference-papers/drs2020/researchpapers/109

Morgan, A., & Barden, M. (2015). *A beautiful constraint: How to transform your limitations into advantages, and why it's everyone's business.* Wiley.

Pye, D. (with Internet Archive). (1971). *The nature and art of workmanship.* New York, Van Nostrand Reinhold. http://archive.org/details/natureartofwor00pyed

Rose, D. B. (2017). Shimmer: When all you love is being trashed. In *Arts of living on a damaged planet: Ghosts and monsters of the anthropocene.* University of Minnesota Press. https://compostingfeminisms.files.wordpress.com/2019/08/shimmer-150.pdf

Roth, G. (1997). *Sweat your prayers: Movement as spiritual practice.* Taschen.

Sinner, A., Hasebe-Ludt, E., & Leggo, C. (2018). Long story short: Encounters with creative nonfiction as methodological provocation. In P. Leavy (Ed.) *Handbook of arts-based research* (pp. 165–189). Guilford Publications. https://www.guilford.com/books/Handbook-of-Arts-Based-Research/Patricia-Leavy/9781462540389

St. Pierre, E., & Pillow, W. S. (Eds). (2000). *Working the ruins: Feminist poststructural theory and methods in education.* Routledge.

Sturrock, G., & Else, P. (Eds). (2007). *Therapeutic playwork reader 1995–2000.* Common Threads Publications Ltd.

Taussig, M. (2012). I'm so angry I made a sign. *Critical Inquiry, 39*(1), 56–88. https://doi.org/10.1086/668050

Taussig, M. (2018). *Mimesis and alterity: A particular history of the senses* (1st ed.). Routledge. https://doi.org/10.4324/9781315227689

Ungley, P. (in press). Put your loving hand out: Manifesting rebellious change for an affective university. In *Navigating academic spaces with disruptive care.* Routledge.

Vance, J. D. (2021, November 2). The universities are the enemy. In *National conservatism conference*, Orlando, FL, USA.

Verger, N. B., Duymedjian, R., Wegener, C., & Glăveanu, V. P. (2024). Creative preservation: A framework of creativity in support of degrowth. *Review of General Psychology, 28*(3), 268–281. https://doi.org/10.1177/10892680241256312

Winnicott, D. W. (1960). The theory of the parent-infant relationship. *The International Journal of Psycho-Analysis, 41*, 585–595.

CHAPTER 9

The Mythologies of Possibility: Global South, Care, and Decolonial Scholarship

Leyya Hoosen and Karabo-Maya Rodwell

Introduction

Our paper emerges from, and as, a continued dialogue we have been having over the last couple of years. As intersectional decolonial scholars, we have chosen to write this chapter in a format that reflects and preserves our engaged, but informal, tone which permeates our conversations. Versions of this meditation have been presented at the African Humanities Association's Annual Conference in 2023 at the University of Cape Town, the Possibility Studies Network's Annual Conference in 2024 at the University of Cambridge, the World Anthropological Union Conference in 2024 in Johannesburg, South Africa, and the keynote we delivered at

L. Hoosen
Department of Political Studies, University of the Witwatersrand, Johannesburg, South Africa
e-mail: leyya.hoosen@wits.ac.za

K.-M. Rodwell (✉)
Department of Anthropology, University of the Witwatersrand, Johannesburg, South Africa
e-mail: karabo.rodwell1@gmail.co.za

N. B. Verger, D. Mattson (eds.), *Possibility Studies and Societal Futures*, Palgrave Advances in Possibility Studies,
https://doi.org/10.1007/978-3-032-26379-7_9

the Possibility Studies Network's Virtual Conference in 2025. The conversation that follows touches on a number of key themes, including discussions of the breakdown in certainties of the everyday in the South African and Global South contexts, considering mythologies of possibilities and their socio-political impacts, and engagements with what it means when we consider possibility being possible. Within this, we draw on the Creative Preservation Framework, which allows us to examine how creativity and imagination operate as survival strategies in challenging circumstances, and how innovation itself may be understood as a form of resistance and care (Verger et al., 2024; 2025)
). In doing so, our discussion foregrounds the generative potential of crisis, how moments of rupture and instability open up conditions for possibility, inviting us to reimagine not only what can be preserved but also what can be created anew. By structuring this piece as dialogic, we are drawing on established traditions that write in a particular way to break down hierarchies in authorship, academic research, and writing practices (Camiciotti, 2014, Chigumadzi (2021), Clifford & Marcus (2023), Fassler (2017), Hooks (1992), Meer (1997), and Nyamnjoh (2012)).

Two key ideas breathe through the following conversation. First, the power of language and articulation. Throughout our work, we have chosen to speak in ways that reflect the truths of our relationships and contexts. Grounded in rigorous (decolonial) academic frameworks, intellectual partnership, and creative practices, we also strive to ensure our language is accessible. This deliberate choice is an intervention in itself, countering the exclusionary tendencies that often characterise academic discourse. In doing so, we affirm our agency as black women, emerging scholars, and contributors to knowledge production, while creating space for others to join the conversation. Secondly, we emphasise the importance of the relationships we have built through this work. Our intellectual partnership has unfolded through countless conversations and is driven by commitments to care, love, and mutual respect. In this spirit, we invite you, dear reader, to join us in this practice. We do not offer our thoughts solely for academic consumption but invite you into the ongoing, relational process of constructing knowledge that is mindful of contexts, grounded in the experiences of and from the Global South, and open to the transformative potential of seeing the Self in the Other (BlackPast, 2012).

Karabo Speaking on the Breakdown in Certainties and Histories

My ideas about possibility, youth, and dreaming in South Africa were cemented after reflections of the 29th of May 2024, which marked South Africa's 7th democratic elections. This was the second time I had cast my vote in national elections, and for many of my contemporaries, it was also the second or third electoral season. The debates that emerged from these elections highlighted for me what possibility truly means in the South African context.

My PhD focuses on the experiences of such youth in post-liberation societies in southern Africa, specifically in Mozambique and South Africa. Many of us experience our nation as a cloud, rather than a "rainbow", as our former President Nelson Mandela termed it. There is a sense of sitting on the edge of liberation. As South Africa is praised globally for our constitution, we're at the pinnacle of morality, yet the everyday reality of South Africans pushes us further and further away from believing in this.

South African youth have a history of resistance. It is important to recognise that the victors of liberation movements do not offer a complete picture of the diverse and divergent narratives that exist across these locales. As Graham states, "for societies undergoing transition such as South Africa, interpretations of the past are extremely important in shaping the present and the future, requiring a shift in thinking about the purpose of history" (Graham, 2014, p. 3). Critiques of liberation movement utopias, such as Graham's, emphasise that (re)writings of history are too often manipulated to advance certain agendas and narratives. This is also evident in the romanticisation of liberation movements, with darker truths sidelined.

Yet, we still argue that South Africa is at the precipice of change. We have gone through cycles of being at the precipice since democracy in 1994, but there is a breakdown of certainties in the everyday which makes us question what we are working towards. However, change is still happening in spaces such as the University of the Witwatersrand (one of the biggest universities on the continent) and the University of Johannesburg (one of the fastest growing universities on the continent). This is the result of possibility being possible. For us, this stems from recognising the Self in the Other and understanding that, in order for possibility to be possible, we need to find ways to work together.

Are we romanticising a failed liberation state? Maybe. We don't know what we are aiming for, even as we work towards it. This is evident in countries throughout the "Global South", from South Africa to Brazil, from India to Argentina, all of whom have endured the weight of colonialism, imperialism, and economic exploitation. When I speak to my Brazilian colleagues and friends, I am constantly struck by how many similarities we share, even though we are thousands of kilometres apart. The historical threads that bind us, from the legacy of colonial domination to the struggle for autonomy and dignity, are undeniable. This shows that the "Global South" is not only geographical but also ideological, historical, and political. And yet, we must ask: How often are we labelled as the Global South without recognition of these shared histories? How often do we see the Global South peripheralised?

The struggles of the Global South, marked by the historical weight of colonialism and ongoing economic exploitation, create an environment in which not only entire nations are at the periphery but so too are our youth. We must note that there is no society without youth dreaming. Without careful consideration, these dreams will disappear. As we do not have space to dream, hope, and think about what is possible when the majority of South Africa is worried about making it to tomorrow, thinking about 20 years from now is too far off.

Leyya Considers the Role of Mythologies of Possibility

Deleuze and Guattari, in their text *What Is Philosophy?* (Deleuze & Guattari, 1996), begin by asking what a concept is. In their unpacking of this question, a line stood out to me. They say, "possible worlds have a long history" (Deleuze & Guattari, 1996, p. 17). This stood out for a few reasons. Firstly, because of the phrase "possible worlds". This could refer to worlds that are possible but yet to be realised, and to worlds that are realised and yet to be experienced. For the purposes of the sentence, it almost doesn't matter which reading it is, because both are equally valid. This sci-fi-esque setting of "possible worlds" that is simultaneously our reality and more-than-our-reality is invoked to make the point that the worlds "have a long history". Again, there is an almost cyclical nature to their writing, because Deleuze and Guattari refer to a "long history" (Deleuze & Guattari, 1996, p. 17). They emphasise the timescape within which these "possible worlds" exist, and it seems to me that this emplacement of possibility within a socio-political history that is "long", entangled, and contested is crucial. What we imagine to be possible is deeply impacted by the myths we hold about possibility.

I define the mythologies of possibility as "the beliefs or stories that we hold about our condition and what might or might not happen for us in our lifetimes". These myths are the narratives we hold about the perceived facts of our existence. The realms -the possible worlds - within which these myths operate are contested, evolving, and political. Possibilities are not neutral spaces, imaginations, or myths. For this reason, there is a particular urgency in critically examining the mythologies of possibility, because the stakes are high: the mythologies we adopt can either reinforce oppressive structures or challenge them.

In South Africa, for example, our positionality - our race, gender, socio-economic status, and how we are positioned within the Global South - significantly shapes the narratives of possibility we deem available to us. It is important to note that when we speak of the Global South, it is as much a geographical reality as it is an ideology, operationalised in numerous ways to shape power dynamics, opportunities, and perceptions. This mythology around the Global South is essential to interrogate. The terminology we use should centre on acknowledging shared histories and interconnected struggles rather than perpetuating an "us" and "them" binary (Spivak, 2023). This shift is ideologically significant, moving towards solidarity and mutual recognition, challenging divisive narratives that often underpin existing mythologies. How society perceives us and how we are read profoundly impacts what we are told is within the realm of possibility.

For many students in South Africa, especially in the context of a young democracy, the dominant mythology tells them two things: they should either be grateful simply to exist within these spaces (the myths of the born-free generation) or strive to be exceptional to justify their presence (the struggle for exceptionalism becomes problematically viewed as a privilege). There is rarely an allowance for the "in-between" - the possibility of being good enough or simply present and engaged. This mythology is inherently limiting. It is imposed, not inherently true, and often operates to sustain systemic inequalities. To return to the quote that opened this consideration, that "possible worlds have a long history" (Deleuze & Guattari, 1996, p. 17), a question begs to be asked whether there is a singular history to speak of or whether we must rather think of what it means that "possible worlds have [long histories]"? By moving further towards multiplicity, considering the dynamic entanglements of histories as they weave across a timescape that then gives birth to possible worlds, we shift our thinking away from the complacency of thinking of the singular and mapping possibility onto a singular realm/plane. Instead, possibilities and the myths we hold about what is possible begin to come into being through being-in-relation with our histories.

Returning once more to Deleuze and Guattari's considerations of the concept, I invoke them, saying that "new concepts must relate to our problems, to our history, and, above all, to our becomings" (Deleuze & Guattari, 1996, p. 28). With the slight alteration of inserting "histories" where they use "history", I want to explore why their framing of the concept is important here. In considering the realm of possible worlds, the myths of possibility that are operationalised in these worlds, and the histories with which they are intrinsically entangled, I resonate with Deleuze and Guattari's impulse to pause and consider what it is we are exploring, defining the problems facing us as we do so, and taking the time to critically unpack the terms and concepts we choose to work with in the, days, decades, and centuries to come. We are becoming in many ways, in who "we" are, who the discipline is and will be, the concepts that will be operationalised and come to be canon, and the list of becoming continues.

Deleuze and Guattari remind us that "mental landscapes do not change haphazardly through the ages: a mountain had to rise here or a river to flow by there again recently for the ground, now dry and flat, to have a particular appearance and texture" (Deleuze & Guattari, 1996, p. 86). If we are to take seriously a discipline that maps new ground, we must build into this practice an interrogation of the landscapes we are re-mapping and turn a critical eye on the mythologies, ontologies, epistemologies, and cosmologies that we are working with. This interrogation of mythologies of possibilities extends into the academy in various ways. Our research methodologies and approaches must seek to be experimental, attentive to difference and nuance, and intentionally stake a claim for a different narrative about what is possible within the academy. The topics we choose to research are not neutral; they carry their own stakes. The act of researching is itself a form of activism - an effort to rewrite the mythology of what scholars like me can and cannot do. Building community within the academy is another essential way we challenge these narratives. Our relationship to each other and to the idea of community is transformative. Community is more than a support system; it is a radical tool for uplifting, sharing opportunities, and creating spaces where none previously existed.

We've decided there is room for both of us. And where there isn't room? We will make it. This applies to lecturing, presenting, and the shared labour of being in academia. There are spaces that necessitate individual effort, but many more demand a communal approach. Working in isolation is not a sustainable future; community is where possibility thrives.

This interrogation of mythologies, this effort to create space, is a political undertaking that affects all aspects of how we conduct and enact scholarship and pedagogy. As black scholars, as members of the Global South, we know how often the deck is stacked against us. Yet, there is also something profoundly exciting about engaging with the possibility of remapping these myths. The question is not only what we want to be possible but also how we go about creating those possibilities, collectively and intentionally.

Karabo Responds, Unpacking Her Views on Possibility Being Possible

As we enter any space, it is crucial to recognise that each of us arrives as a whole person, with our own experiences, ideas, and, importantly, our own care practices. When I speak of care practices, I mean not merely acts of nurturing but vital creative strategies that enable us to engage meaningfully in the work we do together. In building a community, it is not about starting from scratch or pretending we are all blank slates. Instead, we come with our foundations already laid, shaped by our personal histories, struggles, and ambitions that inform how we contribute.

It is important to acknowledge that you, I, and anyone else entering this community come as full people with thoughts, ideas, and our own care practices. When I speak of care practices here, I mean them as crucial creative strategies. The idea of building rather than creating is that everyone has a foundation, and now we are putting this together and seeing what comes from it, rather than pretending we are all coming into this as blank slates. As a colleague in my department often says to her students, "Do not leave yourself at the door when you walk into a lecture." This resonates deeply with me, as I see it apply not just to the classroom but to all spaces we inhabit, including the academy, our communities, and beyond. When we enter these spaces, we bring with us the entirety of who we are. And that's precisely the point: we are not merely filling roles; we are engaging with the world through the lens of our lived experiences, the struggles of our histories, and the contexts that shape us.

Building community isn't easy, but it is necessary for survival, for thriving. This is where radical care practices come in. We cannot simply push forward alone; we need each other to heal, grow, and survive in systems stacked against us. The more sure I became of my path, the more I understood the need to be part of a community that nurtures and supports that growth. When we met, it felt like an opportunity to learn from someone already ahead, someone who could show me not only how to avoid mistakes but also how to recover from them when they inevitably happen.

I am reminded of discussions around the "Master-Apprentice" relationships that often dominate our academic structures. But as we've worked together, we've moved beyond this hierarchical, top-down approach to knowledge sharing. Instead, we have intentionally always engaged each other as equals, sharing our thoughts and experiences freely, recognising that we all have something valuable to contribute. It is in this egalitarian, collaborative approach that we find our strength and resilience. In a world that often devalues our knowledge and contributions, this practice of mutual care is not only radical; it is essential to our survival and success.

Here, we need to make a crucial distinction between knowledge production and knowledge (re)production. For too long, scholars in the Global South have been relegated to reproducing knowledge created elsewhere, often shaped by frameworks that do not reflect our realities. Knowledge production, however, is about creating new understandings, challenging dominant narratives, and asserting our agency in defining what knowledge is and who gets to create it. This is especially vital for us in countries and spaces where our histories and innovations have been overlooked. We must actively reshape systems of knowledge to reflect our lived experiences and futures. Equally important is the inclusion of young people in this process. As emerging scholars, we must not only consume knowledge but also contribute to its creation, ensuring a more inclusive and dynamic academic landscape that reflects the voices and contributions of the Global South. Many of us are actively doing this by producing new knowledge, with out wording, our histories, ourframing. I would argue that these ongoing cycles of learning demonstrate our commitment to theory-to-practice cycles of knowledge.

This echoes much of the work of theorists such as Edward Said, who argued that Western epistemologies have historically defined what counts as legitimate knowledge, often marginalising and distorting the lived experiences and intellectual traditions of the Global South (Said, 1994). Frantz Fanon's The Wretched of the Earth (Fanon, 1967) and Gayatri Spivak's (Spivak, 2023) work on subalternity question the structures that deny agency to the oppressed, calling for a reclamation of the power to define and create knowledge.

As we have said, in our work we have often followed a process that feels intuitive. We talk through a chapter, section or presentation while the other person writes it down, back and forth, constantly refining our ideas together. This practice has become second nature to us, a default. But as we all sit here, writing in this academic space, I find myself questioning: Is this the kind of thing weare meant to discuss in an academic book? Is it even academic enough? Yet, after reflecting on it, I've come to a realisation: this process, the collaboration and exchange, is precisely what carries us through academia. It is what has allowed us to navigate the Masters, PhD, and employment journeys. It may not fit neatly into traditional academic moulds, but it is absolutely central to our scholarship and to how we have persevered.

This brings me to a final set of questions, which I think are vital for all of us to reflect on. Have we simply "made this up", creating a support system that doesn't exist in traditional academic spaces? Or are others also doing this, even if we don't always see it? I know many young scholars are working alone, navigating this path without the support of a community. So I wonder: Have we romanticised the academic experience? Or, more importantly, are we challenging the very systems that have historically excluded young people, women, and black people from meaningful participation in knowledge production?

It is essential, especially for us as young scholars, to recognise the role of collaboration in how we engage with knowledge. Too often, academic structures do not know what to do with us; they do not know how to engage with us, they do not actively include us, nor do they provide spaces for us to shape the very knowledge systems we are part of. This is why we must continue to reshape these systems, moving beyond traditional academic structures that isolate rather than include. True knowledge production must be a collective, inclusive process that allows young voices to be heard, not just as recipients of knowledge but as creators of it. We are the ones who will define the future of scholarship, but only if we insist on making space for ourselves and for each other.

In doing so, we can hold ourselves accountable. Accountability is not only about checking whether we practise what we preach about radical care; it is about ensuring our actions align with our words and that we create systems that support us and others. The care we offer each other in our academic journeys is an essential survival strategy. It is how we thrive, how we persist, and how we push back against systems that seek to marginalise and exclude us.

Leyya Offers Her Views on Possibility Being Possible

Permit me a long quote to begin my response. Harney and Moten, in their text on the undercommons and the role of hapticality within it, say:

"Never being on the right side of the Atlantic is an unsettled feeling, the feeling of a thing that unsettles with others. It's a feeling, if you ride with it, that produces a certain distance from the settled, from those who determine themselves in space and time, who locate themselves in a determined history ... Hapticality, the touch of the undercommons, the interiority of sentiment, the feel that what is to come is here. Hapticality, the capacity to feel through others, for others to feel through you, for you to feel them feeling you ... the invitation to build sentimentality together again, feeling each other again, how we party. This is our hapticality, our love" (Harney & Moten, 2013, p. 98).

This idea of hapticality, or love, is the crux of remapping mythologies of possibility. Whether it's Harney and Moten's hapticality (Harney & Moten, 2013), Deleuze and Guattari's conceptualisation of the Friend (Deleuze & Guattari, 1996), Tuck and Yang's refusal politics (Tuck & Yang, 2014), Ahmed's queering of orientation (Ahmed, 2006), Spivak's subaltern (Spivak, 2023), Alexie's reservation of the mind (Alexie, 2017), Walker's artist and mother's garden (Walker, 2000), or even our (Hoosen and Rodwell's) theorising of theory-to-practice - all these are invocations of love, of connection, and a deep sense of empathy. These are calls for types of scholarship and pedagogical practices that are themselves haptic, that are themselves loving.

This ethos has been central throughout Karabo-Maya's and my own journey as we have navigated various phases and identities while being in-community with each other. We found our scholarly and pedagogical journeys overlapping, collapsing into one another, separating out, and collapsing again and again. These journeys are cyclical - and so are the problems, gazes, and complications that come with them. We have been students together, sharing what scholarship means to us as students undertaking degrees and researching, continuously learning and defining scholarship in experimental ways. We have been lecturers together, figuring out what pedagogy means to us - creating experimental and impactful ways to link theory to practice, bringing our whole selves into the classroom and having students see that, to remind them that they too belong and matter in these spaces. Academia isn't just for a privileged few - it can reflect the diversity of those teaching and lecturing - it is an act of care and transformation. We have, throughout our individual and collective journeys, navigated the gaze of the academy. We've experienced the constant and unrelenting gaze of the academy - a gaze that both scrutinises and shapes. Together, we've found ways to negotiate visibility, opacity, and accountability to make possible what we strive for. Bouncing ideas off one another and holding each other accountable is not only practical but necessary for joy, resilience, and success in academic life.

This haptic process, these journeys of love, have allowed us to continuously reflect, revise, and redefine what success and fulfilment mean for us. It has also underscored the difficulty of finding and sustaining community in academia, but once we built a springboard between the two of us, it became easier to welcome others into this space.

These cycles of community and collaboration - building theory, practising it, and constantly learning how to make spaces more accessible and inclusive - are intentional acts of resistance. These cycles of practice remind us that silos may be common in academia, but they are not inevitable. Reciprocity ensures that even in challenging times, we hold each other accountable in the healthiest ways.

In doing all this, we see the undercurrent of the academy's Gaze (hooks, 2012) - ever-present and strategic in shaping how we navigate what remains unsaid. This tension is central to our praxis: the visible and the unspoken coexisting, reflecting the complex strategies of working within and against institutional norms.

We have been having this conversation for years, presenting different iterations of this paper at conferences across continents and speaking about this topic in a range of spaces. The necessary work of survival - both in academia and in life more broadly - requires thoughtful, deep care. This is a conversation that must continue; there isn't a singular answer, as our needs evolve over time and our care practices change as we grow. Engaging in these discussions intentionally and compassionately, weaving them through our scholarship, allows us to collectively imagine possibilities we have yet to think of.

Leyya and Karabo-Maya Conclude (for Now)

Reflecting on our journey, we find ourselves in a continuous process of self-reflection and recognition. Our understanding of strategies of the self has evolved, increasingly shaped through the lens of the Other. This journey, moving from self-actualisation to seeing the Self in Others, resonates with Audre Lorde's powerful assertion that self-definition is essential (BlackPast, 2012). This act of self-definition, we argue, is inseparable from the recognision of the interconnectedness of individual and collective identities. We bring to the table not only our individual perspectives but also the shared histories and struggles that have shaped us.

In considering the South African academy within the broader context of the Global South, we must acknowledge the limits of transposing models of possibility from the Global North. The realities we face here are deeply informed by histories of colonialism, apartheid, and socio-political struggles that continue to shape our contemporary lives. These histories have impacted the socio-political fabric of the Global South, histories that cannot be simply understood through the paradigms of the Global North. The work we are engaged in, therefore, demands a nuanced approach that accounts for these complexities while engaging processes of transformation that are rooted in the distinct realities of the Global South.

Two key ideas remain central as we conclude this chapter. First, the power of language and articulation. Throughout our work, we have purposefully chosen to speak in ways that reflect the truths of our relationships and contexts. While we have remained grounded in rigorous academic frameworks, particularly those concerned with decoloniality, intellectual partnership, and creative practices, we have also striven to ensure that our language is accessible. This deliberate choice is an intervention in itself, countering the exclusionary tendencies that often characterise academic discourse. In doing so, we affirm our agency as black women, emerging scholars, and contributors to knowledge production, while creating space for others to join the conversation. Second, the importance of the relationships we have built through this work. Our intellectual partnership has unfolded over countless conversations, driven by a commitment to care, love, and mutual respect. It is in this spirit that we invite you, reader, to join us in this practice. We do not simply offer our thoughts for academic consumption but invite you into the ongoing, relational process of constructing knowledge that is mindful of context, grounded in the experiences of the Global South, and open to the transformative potential of seeing the Self in the Other.

In the Global South, possibility is not a distant ideal but a lived reality, shaped by the resilience and creativity of those who continue to resist and imagine new futures. It is in this space of possibility that we find hope, not only for our own futures but for the futures of our communities, our countries, and our world. The journey from self-actualisation to recognising the Self in Others is an essential step in this ongoing work.

References

Ahmed, S. (2006). Orientations: Toward a queer phenomenology. *GLQ: A Journal of Lesbian and Gay Studies, 12*(4), 543–574.

Alexie, S. (2017). The reservation of the mind. In J. Fassler (Ed.), Light the dark: Writers on creativity, inspiration, and the artistic process (pp. 23–31). Penguin Books.

BlackPast. (2012, August 12). Audre Lorde: "Learning from the 60s." https://www.blackpast.org/african-americanhistory/1982-audre-lorde-learning-60s/

Camiciotti, G. D. L. (2014). Letters and letter writing in early modern culture: An introduction. *Journal of Early Modern Studies, 3*, 17–35.

Chigumadzi, P. (2021). Hearing the silence. In D. Lewis & G. Baderoon (Eds.), *Surfacing: On being black and feminist in South Africa* (pp. 226–237). Wits University Press.

Clifford, J., & Marcus, G. E. (Eds.). (2023). *Writing culture: The poetics and politics of ethnography*. University of California Press.

Deleuze, G., & Guattari, F. (1996). *What is philosophy?* Columbia University Press.

Fanon, F. (1967). *The wretched of the earth* (C. Farrington, Trans.). Grove Press.

Fassler, J. (Ed.). (2017). *Light the dark: Writers on creativity, inspiration, and the artistic process*. Penguin.

Graham, M. (2014). The ANC and the "myth" of liberation solidarity. *Africa Insight, 44*(1), 1–15.

Harney, S., & Moten, F. (2013). *The undercommons: Fugitive planning and black study*. Minor Compositions.

Hooks, b. (1992). *Yearning: Race, gender, and cultural politics*. South End Press.

Hooks, b. (2012). The oppositional gaze: Black female spectators. In M. Diawara (Ed.), *Black American cinema* (pp. 288–302). Routledge.

Meer, S. (1997). Feminist agendas in activism and academia: The challenge for agenda. *Agenda, 13*(34), 5–16.

Nyamnjoh, F. B. (2012). Blinded by sight: Divining the future of anthropology in Africa. *Africa Spectrum, 47*(2–3), 63–92.
Said, E. W. (1994). *Orientalism*. Vintage Books.
Spivak, G. C. (2023). Can the subaltern speak? In *Imperialism* (pp. 171–219). Routledge.
Tuck, E., & Yang, K. W. (2014). R-words: Refusing research. In D. Paris & M. T. Winn (Eds.), *Humanizing research: Decolonizing qualitative inquiry with youth and communities* (pp. 223–248). SAGE.
Verger, N. B., Duymedjian, R., & Glăveanu, V. P. (2025). The meritocracy of preservation: Reimagining merit beyond production. *Organization, 33*(2), 294–312.
Verger, N. B., Duymedjian, R., Wegener, C., & Glăveanu, V. P. (2024). Creative preservation: A framework of creativity in support of degrowth. *Review of General Psychology, 28*(3), 268.
Walker, A. (2000). In search of our mothers' gardens. In J. D. Garner (Ed.), *Worlds of difference: Inequality in the aging experience* (p. 48). Pine Forge Press.

CHAPTER 10

The Vitality of 'Stuporism': Advancing a New Conceptualisation for 'Transformative Scientific' Inquiry and Educational Innovation

Antonia Symeonidou, Danilo Audiello, and Pamela Burnard

Introduction

Imagine these choreographies of possibility:

Beneath the ocean's surface, coral reefs are quietly building underwater cities—vast, intricate architectures constructed by tiny polyps no bigger than a grain of rice. These living structures don't just exist; they regulate marine ecosystems, shelter 25% of all ocean life, and even dissipate wave energy to protect coastal communities (Moberg & Folke, 1999). But here's the marvel: they're **alive**, fragile, and fiercely cooperative.

A. Symeonidou • D. Audiello
Academy of Magic & Science, University of Cambridge, Cambridge, UK
e-mail: danilo.audiello@unifg.it

P. Burnard (✉)
Faculty of Education, University of Cambridge, Cambridge, UK
e-mail: pab61@cam.ac.uk

N. B. Verger, D. Mattson (eds.), *Possibility Studies and Societal Futures*, Palgrave Advances in Possibility Studies,
https://doi.org/10.1007/978-3-032-26379-7_10

But *how* is that possible?

Now consider the forest floor. Underneath our feet, a vast fungal network—*mycelium*—is running messages, redirecting nutrients, and balancing entire ecosystems. Research shows that trees send distress signals through these networks, and neighbouring trees respond with biochemical support (Simard et al., 1997; Sheldrake, 2021). Certain fungi have even been found digesting **radiation** at the Chernobyl site (Dadachova et al., 2007), and others are being studied for their ability to break down **microplastics** (Sánchez, 2020).

But *how* is that possible?

Shifting our gaze to an ant colony. No leader gives commands, yet thousands of individuals build tunnels, forage food, and adapt their behaviour instantly to environmental shifts. These decentralised systems exhibit problem-solving, memory, and task allocation that rivals algorithmic efficiency (Gordon, 2010).

But *how* is that possible?

We blink. A butterfly emerges from its chrysalis, having liquefied itself entirely during metamorphosis, rearranged its molecules, and become something new. Scientists have found that certain structural elements and even memories survive this process (Blackiston et al., 2008). No plan, no prototype—just nature's astonishing choreography of transformation.

But *how* is that possible?

These are not fairy tales or metaphors designed to inspire wonder alone. They are real, observable phenomena that confront observers with moments of radical unexpectedness—moments in which existing explanatory frames fail, an assemblage of scenarios of **'stuporism' in action**! Encounters that generate a sudden epistemic rupture, **short-circuiting our assumptions**. These moments do something profound. They don't just inspire, they demand explanation, compelling the question "How is this possible?" (Audiello & Symeonidou, 2024). They pull us, wide-eyed, into the territory of science. This emphasis on questioning resonates with Heidegger's insistence on the fundamental significance of the question itself (Heidegger, 1962). For Heidegger, questioning is not merely a methodological step towards answers, but an ontological orientation: a way of opening a field of intelligibility in which phenomena can show themselves as meaningful. A genuine question, in this sense, does not simply seek information; it discloses what matters, shapes the horizon of inquiry, and determines what can be encountered as relevant or possible. Within the present framework, moments of stupor initiate precisely this kind of questioning, not as problem-solving in search of closure, but as an opening that reconfigures how learners relate to phenomena, possibilities, and themselves.

And this is where it begins: **curiosity ignites**, questions multiply, and inquiry becomes inevitable. These flashes of awe are not the end of wonder—they are its invitation. What follows is the long, often messy, deeply human process of trying to understand. Stuporism does not replace curiosity or wonder, nor does it subsume them. Rather, it marks a temporal and phenomenological threshold, the instant at which sense-making breaks down before reflective engagement begins. In this way, stuporism functions as a catalyst: it opens a space in which curiosity can be activated, and wonder can be sustained. And while wonder is often described as a prolonged orientation of openness and attentiveness, stuporism names the fleeting disruption that makes such an orientation possible in the first place.

From this rupture, inquiry becomes inevitable (see also Mattson, Chap. 11, this volume). Curiosity ignited as learners seek to resolve what has unsettled them, while wonder enables that to remain with uncertainty, rather than foreclose it prematurely. In this chapter, we will explore how **scientific inquiry**, when fuelled by **curiosity**, **wonder**, and especially **'stuporism'—the momentary emotional surge of surprise that becomes a catalyst for deeper scientific inquiry**—does more than illuminate the natural world. It changes how we see society—and what we believe is possible within it. From classrooms to communities, from laboratories to policies, the sense of "how is that possible?" can evolve into "what might we do with this understanding?"

In an age marked by uncertainty, fragmentation, and transformation, perhaps the most radical and rebellious thing we can do is **teach awe**. To give learners not just tools and facts, but experiences that disorient just enough to open new pathways of thought. Because sometimes, to truly change the world, we might just first need to be astonished by it!

There are known knowns, things we know that we know; and there are known unknowns, things that we know we don't know. But there are also unknown unknowns, things we do not know we don't know. Some moments in human history, science communication, or simple observation of everyday phenomena, don't just add to what we know. They disrupt the very assumptions that organise knowledge itself. Such moments re-position learners within a state of productive epistemic vulnerability, a condition in which curiosity becomes more than a method, and wonder becomes more than a luxury. They become necessary orientations for navigating complexity, drawing on different forms of thinking to understand, beyond a single epistemic source or method, thereby inviting transformation and change.

Consider these practices of 'stuporism':

In 1895, German physicist Wilhelm Röntgen, the first recipient of the Nobel Prize in Physics in 1901 and known today for X-rays (or Röntgen rays), discovered something no one had ever seen before—by seeing through it. Wilhelm Röntgen was the first to systematically produce and detect electromagnetic radiation at a wavelength unknown until then. A mysterious, invisible ray could pass through flesh and reveal bone. The first X-ray image was of his wife's hand; she looked at it and said, "I have seen my death". It was beautiful and terrifying. The world had changed—the invisible was now visible (Glasser, 1993).

What realities are hiding in plain sight—or out of sight?

In a small elementary school classroom, a teacher sets up a simple physics demonstration: two metal spheres, one heated, the other not. The students watch the hot sphere expand and no longer fit through the ring it had just passed through seconds earlier. Their faces shift—from confusion to astonishment to inquiry. That moment of surprise—the "what just happened?"—is a pedagogical doorway. A child raises their hand: "Does this mean things are always moving, even when we can't see it?" Such classroom wonder is not trivial—it's a scientifically productive moment of epistemic disruption (Keltner & Haidt, 2003; Schinkel 2017).

If this is possible, what else can there be?

In 1943, abstract painter Jackson Pollock began working not with brushes, but with movement. He walked around the canvas, flung paint, bent space into time. When critics saw the results, they didn't call it "expression". They called it "chaos". But **within the chaos, mathematicians later saw fractals**—self-similar patterns now known to structure coastlines, lightning, and even lung tissue (Taylor et al., 1999).

What does this reveal about the limits of what we thought we knew?

In 1990, the Voyager 1 spacecraft, already more than 6 billion kilometres away, turned its camera back towards Earth and took a photograph. The result was a tiny dot suspended in a sunbeam. That pale blue dot, redefined our place in the cosmos—not as rulers, but as temporary passengers on a fragile planet. It was a scientific image, yet it carried the moral weight of poetry.

What other assumptions are we carrying without realising it?

At the Large Hadron Collider in 2012, after decades of theorising, scientists finally observed the Higgs boson—the elusive "God particle" responsible for giving mass to matter (Greene, 2004). It had been there all along, invisible and undetectable, yet fundamental to reality. The discovery wasn't just a confirmation. It was a reminder that **what we can't see still shapes everything we are**.

What other truths have we yet to see?

It was with this question in mind that Danilo, one of the co-authors, visited the new facilities at CERN, exploring their scientific frontiers through a science-outreach interview and stuporistic storytelling for the project Dante in Puglia celebrating the 700th anniversary of Dante Alighieri's death. In his narrative, he drew an evocative analogy with Dante's Paradiso, the third chapter of the Divine Comedy's journey through Hell, Purgatory, and Paradise. Just as Dante's ascent reveals increasingly radiant realms of knowledge, CERN's newest experiments, conducted in its new facilities, aim to produce more particle collisions per second, generating what physicists describe as a "brighter" beam. Here, brightness refers not to visible light intensity but to the unprecedented density of collisions, aiming to push the frontiers of science by exploring unknown physics such as dark matter, extra dimensions, and Higgs boson properties. Through this storytelling lens, the work at CERN becomes more than an engineering feat; it becomes an invitation to reimagine what lies beyond the boundaries of human perception and to confront the limits of what we can see.

These moments—across classrooms, laboratories, studios, and space—are instances of 'stuporism' in action. They are fleeting, when surprise becomes an epistemic catalyst, igniting new pathways of scientific inquiry. They are the moments that make the world feel new again, not by giving us answers but by undoing our certainty. They return us to a childlike posture of awe—one that says not "I understand", but "I am open".

While these moments are described here as instances of stuporism, they do not exhaust the experiential arc that follows such encounters. Recent work in possibility and creativity studies emphasises that sustained engagement with uncertainty depends on affective orientations that extend beyond the initial rupture. Verger (2024) argues that encounters with possibility are sustained not by cognition alone but through affective engagement that allows individuals to remain attuned to what could be, even when outcomes remain indeterminate. From this perspective, stuporism can be understood as the initiating epistemic rupture, while wonder functions as the sustained aesthetic–affective orientation that stabilises openness and enables inquiry to unfold over time. This is why 'stuporism' belongs at the centre of education. Education is not just about acquiring facts—it's about learning how to be engaged in transformation. It's about raising a generation who can feel that same astonishment—and then lean into it, asking: If this is possible, what else can there be? And that question, the one that begins in wonder and ends in action, may be the most radical tool we have for reimagining science, society, and the future itself!

From this perspective, the six-step framework we propose below should not be read as promoting stuporism in isolation. Instead, it models how designed moments of epistemic rupture can initiate longer trajectories in which curiosity and wonder alternately activate, sustain, and deepen inquiry.

The six-step framework proposed in this chapter was developed through iterative synthesis of multiple strands of empirical, theoretical, and design-based work by the authors over time. These strands include research in science and arts education, inquiry-based learning, and practice-led experimentation with illusion and performance as epistemic tools. Across these contexts, a recurring pattern emerged: moments of surprise or epistemic disruption heightened engagement, questioning, and collaborative inquiry. When such moments were given time for questioning and "what-ifs" rather than being resolved too quickly, they often led to deeper exploration and speculative thinking. The framework, therefore, brings together what was repeatedly observed across classrooms, public engagement settings, and creative inquiry environments. Importantly, the framework is not presented as a rigid sequence or prescriptive pathway. While it follows a typical progression from encounter to enactment, its phases are intended to be understood as dynamically interacting and context-sensitive. Learners may move back and forth between curiosity and wonder, or re-enter states of stupor as new disruptions arise. The value of the framework lies not in its linearity but in its capacity to make visible the conditions under which inquiry deepens and transformation becomes possible.

To map this process, we invite you to consider this six-step framework for how scientific inquiry, when fuelled by stupor, curiosity, and wonder, can catalyse societal change (see Fig. 10.1). This model draws on cognitive science, education, performance magic, and artistic perception, where illusion, ambiguity, and symbolic disruption often lead to new ways of knowing, seeing, and being. This adaptive model is yours to build upon, apply in your domain, and transform into new practices of teaching,

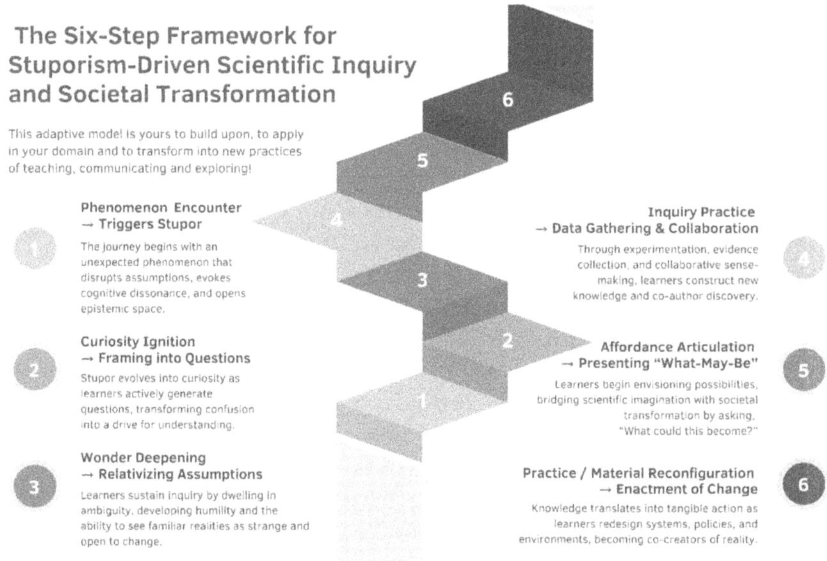

Fig. 10.1 The six-step framework for 'stuporism'-driven scientific inquiry and societal transformation

communicating, and exploring, whether in education, science, policy, design, or the arts.

Phases for/of Adaptive Practice: The Six-Step Framework for 'Stuporism'-Driven Scientific Inquiry and Societal Transformation

Phenomenon Encounter → Triggers Stupor

The journey begins with an encounter: a startling event, an artwork, or a phenomenon that disrupts existing categories and commands attention. Experiencing such epistemic disruption does not depend on innate traits or exceptional sensitivity. Instead, it emerges from a cultivated disposition towards openness, attentiveness, and epistemic humility. These dispositions are not preconditions located solely within the learner; they can be actively engineered by educational and cultural environments that allow and invite uncertainty, slow interpretation, and frame surprise as productive rather than threatening.

Like a magician's illusion, it presents a paradox: what you see contradicts what you know (Kuhn et al., 2008). An artistic parallel is René Magritte's The Treachery of Images, which shows a pipe beneath the phrase "Ceci n'est pas une pipe" (this is not a pipe), disturbing our assumptions about symbols and reality, evoking cognitive dissonance, and opening epistemological space. In science, this moment is echoed in the shock of the unexpected result. In education, it might come from a story, a lab, or a painting that stops students in their tracks. These are aesthetic ruptures, experiences that trigger epistemic stupor (Schinkel, 2017). A musical example is the opening bars of Beethoven's Eroica Symphony, where the second unexpected chord throws off classical listeners. It doesn't belong, yet it opens an entirely new harmonic language, one that signals the transformation of Western music. In this sense, stuporism functions as the experiential micro-event within a broader eruptive logic, where disruption initiates new trajectories of inquiry, participation, and meaning-making, as previously articulated in eruptive research (Symeonidou et al., 2025).

Curiosity Ignition → Framing into Questions

Once stunned, the learner leans forward, implicitly or explicitly asking, "How did that happen?" Curiosity, sparked by stupor, begins to take shape as questions (Loewenstein, 1994). This is not passive confusion; it's the active desire to understand. Importantly, this phase does not presuppose pre-existing intrinsic motivation. Rather, curiosity is understood here as *situationally activated*. Well-designed moments of epistemic stupor can temporarily override disengagement, apathy, or habitual resistance, allowing interest to emerge even among learners who may not initially identify as curious or motivated. A literary illustration appears in Mary Shelley's Frankenstein, also known as *The Modern Prometheus*, the story of a scientist who assembles a creature from body parts and brings it to life. Victor is awed by lightning splitting a tree: "So soon the mind loses its impression of solemnity". This single event ignites his quest to understand nature's secrets. In educational design, this phase is cultivated through open-ended dialogue and environments that reward risk-taking and ambiguity (Rothstein & Santana, 2011). Demonstrations of physics experiments are a prime example of this, as when students witness a magnetic levitation demonstration. These experiments often begin with disbelief but quickly spiral into hypothesis formulation. Their phrasing shifts from "That's weird" to "What if the magnetic field is compensating for gravity?" This is an ignited leap from shock to science.

Wonder Deepening → Relativising Assumptions

Curiosity opens inquiry, and wonder sustains it. Whereas stupor marks the moment when habitual sense-making collapses, wonder describes the learner's capacity to remain with what has been disrupted without demanding immediate resolution. In this phase, learners shift from seeking quick answers to dwelling within questions. Wonder is not set apart from stupor but is what becomes possible after stupor. If stupor interrupts meaning, wonder holds open the space created by that interruption. The two experiences are therefore deeply interconnected yet functionally distinct: stupor destabilises assumptions, while wonder enables those assumptions to be examined, relativised, and reimagined. At this stage, learners shift from seeking quick answers to dwelling in mystery, asking "What assumptions am I bringing to this?" and "What other ways of seeing might be possible?" Wonder invites perspective-taking, humility, and the ability to relativise one's assumptions (Desmond, 2016; Opdal, 2001). A parallel in the world of art is Caspar David Friedrich's Wanderer Above the Sea of Fog, which depicts a figure staring into clouds and mountains, not conquering them but observing and surrendering to their immensity. This could be the aesthetic of wonder: surrender, not mastery!

A theatrical example is Bertolt Brecht's Verfremdungseffekt—the alienation effect—which challenges audiences to pause and see the familiar as strange (Brecht, 1964), thereby provoking critical reflection on content and form. Brecht believed theatre should not lead viewers into passive identification with characters or narratives. Instead, he wanted audiences to see social reality as constructed, changeable, and therefore open to transformation. To achieve this, he used the alienation effect—breaking the illusion of realism through techniques such as actors stepping out of character to address the audience directly, using signs, placards, or projections during scenes, exposing stage mechanics, or introducing abrupt scene changes and stylised movement or narration. The goal was not to move the audience emotionally but to jolt them into awareness. In other words: to create a cognitive rupture—a moment akin to stupor—that makes the familiar seem strange again. His plays halt the narrative to insert reflection, challenging viewers to examine the structures behind reality. Beyond art and theatre, wonder in science, too, delays closure. It invites learners to sit with the unresolved and imagine multiple truths or possible frameworks. This aligns with Mezirow's (2015) concept of the disorienting dilemma, essential for transformative learning. According to Mezirow, a disorienting dilemma is a life event or learning encounter that challenges a person's previously held beliefs, values, or assumptions—causing them to re-examine their worldview. This experience is often emotionally disruptive, cognitively unsettling, and potentially transformative. The learner is faced with something that doesn't fit—that causes epistemic rupture. Like a magician's trick or Brecht's alienation effect, it creates space for critical reflection rather than passive absorption.

Inquiry Practice → Data Gathering, Collaboration

Having destabilised old assumptions, learners now engage in active investigation—collecting evidence, collaborating, and designing experiments. Inquiry becomes the means of making sense of wonder (Hmelo-Silver et al., 2007). Olafur Eliasson's Weather Project, installed at Tate Modern, offers a powerful example of this concept. The artist flooded the vast Turbine Hall with artificial sunlight and fog, creating a surreal, immersive environment. Visitors, confused and mesmerised, began informal, crowd-sourced investigation—testing angles, examining shadows, and theorising the hidden lighting mechanisms. Art becomes collective inquiry. This precise element of co-investigation, driven by awe and uncertainty, can nurture a similarly collaborative spirit in educational settings.

In classrooms, scientific thinking emerges as learners test models, question each other's assumptions, and interpret ambiguous phenomena together. These collaborations simulate the co-constructed knowledge practices of scientific and civic communities (Bonney et al., 2009). Another example is the large-scale citizen science initiative Galaxy Zoo, in which tens of thousands of volunteers help classify images of galaxies taken by space telescopes. Initially designed as a research support tool, the project evolved into a platform for public scientific engagement, where ordinary people began making significant contributions—spotting rare formations and patterns overlooked by algorithms. The collaborative, open-ended nature of Galaxy Zoo mirrors the values of inquiry-based learning: it democratises science, encourages observational rigour, and invites participants to become co-authors of discovery (Lintott et al., 2008).

Whether through the speculative fog of an art installation or the expansive sweep of the cosmos, these inquiry practices reaffirm that curiosity thrives when it is shared, and that wonder deepens when it is scaffolded by evidence, dialogue, and collective exploration.

Affordance Articulation → Presenting "What-May-be"

As learners explore, they begin to articulate new affordances—possibilities that were previously invisible or unimaginable. Affordances, defined as what a given system or tool allows or makes possible, are not fixed but are perceived and interpreted within sociocultural contexts (Gibson, 1979; Schatzki, 2019). To clarify how such affordances come to be perceived through moments of epistemic disruption, Verger (2025) argues that experiences that unsettle habitual frames of reference can expand what individuals perceive as possible, not merely at a cognitive level but through shifts in affective and embodied engagement with the world. In this sense, affordances are not simply discovered as static features of an environment; they emerge through transformative encounters that reconfigure how individuals relate to their surroundings, their capacities, and their sense of agency.

This account resonates with the role of stuporism described earlier, where epistemic rupture opens perceptual and imaginative space for new possibilities to be articulated. Students no longer ask, "What is this?" but "What can this become?" In Italo Calvino's Invisible Cities, each city Marco Polo describes is not a geographic fact but a symbolic possibility, a speculative "what may be?" that reframes how we think about human settlement, memory, and society. This moment of affordance articulation is closely tied to earlier epistemic ruptures. Having experienced stupor and sustained wonder, learners become capable of perceiving alternatives that were previously inconceivable within their pre-existing frames. In this sense, affordances emerge through a process of accommodation, where existing schemes are reconfigured to include new possibilities for thought and action.

This speculative reframing is particularly evident in contemporary scientific breakthroughs such as CRISPR gene editing. CRISPR (Clustered Regularly Interspaced Short Palindromic Repeats) is a revolutionary biotechnology that enables precise, efficient DNA editing—essentially allowing scientists to 'cut' and 'paste' genetic material within living organisms (Jinek et al., 2012). Initially discovered as part of a bacterial immune system, CRISPR has rapidly become a foundational tool in genetics, medicine, and synthetic biology. When students learn about this technology—not just its mechanism but also its implications—they begin to move beyond factual understanding towards imaginative speculation. What if we could cure genetic diseases such as sickle cell anaemia? What if we could engineer crops to thrive in drought-stricken regions? What if we could redesign ecosystems to withstand climate change?

At this stage, inquiry evolves into speculative design—a forward-thinking process that explores not only what is scientifically possible but also what is socially, ethically, and environmentally desirable (Ratti & Claudel, 2016). Students are no longer just asking "how does CRISPR work?" but "what should we do with it?" This shift marks a critical pivot in the learning process: from understanding the world as it is to imagining the world as it could be. Here, the articulation of affordances becomes central to linking inquiry with societal transformation. By identifying what a given phenomenon allows, invites, or constrains, learners bridge scientific imagination with civic foresight. They begin to envision interventions, not in abstraction, but grounded in existing conditions, constraints, and collective values. In education, when students articulate the potential uses and consequences of scientific innovations, they effectively bridge scientific imagination and societal transformation. They begin to prototype futures—not in a utopian vacuum, but through grounded reflection on current conditions, constraints, and values.

Thus, the evolutionary step at the intersection of inquiry and design is not just technical creativity—it is ethical speculation. It is about dreaming responsibly, planning collaboratively, and building prototypes—material, social, and/or narrative—that test what 'could be'. In this way, scientific inquiry becomes an engine for civic foresight and transformative action, turning knowledge into possibility and possibility into agency.

Practice/Material Reconfiguration → Enactment of Change

The final step is material enactment. This may include shifts in personal belief that reconfigure how individuals perceive problems, resources, and agency, as well as more visibly material interventions such as redesigning policies, inventing new products, or reimagining curricula. It is the stage where learning becomes doing—where insights gained through curiosity, wonder, and inquiry are translated into tangible transformations (Freire, 1994).

One iconic example of reconfiguration through artistic intervention is John Cage's 4′33″. In this piece, the pianist makes no sound. While there is a formal score, it contains no musical notes. Instead, the ambient noise of the audience—breaths, coughs, shifting bodies—becomes the music. The audience, by accident or design, becomes the performer. Cage, of course, followed a sheet of music composed of pauses and silences, yet from the audience's perspective it was an unexpected "nothing" where they expected notes. In doing so, he challenged prevailing assumptions about what music is, not by arguing but by staging an encounter that reconfigured perceptual norms. After 4′33″, many never listened to silence the same way again. Similarly, those who have undergone this kind of material reconfiguration may never perceive a social or scientific "problem" as a static obstacle again—but as a dynamic site for transformation.

A pedagogical case that exemplifies this transformation is the "Seeds of Change" project, in which middle school students in a flood-prone urban district turned lessons on climate change into a living intervention (for a discussion of social innovation and creative preservation, see also Verger, Chap. 2, this volume). Rather than write reports or build models, they co-designed and implemented a green infrastructure system on their school grounds. This included rain gardens, permeable paving, and native plant landscaping—all intended to reduce stormwater runoff. What began as a science unit became a civic act. Inquiry became intervention, and a complex issue became a shared challenge—fuelling motivation, ownership, and social-emotional learning (Krasny & Tidball, 2012).

A parallel case in public health innovation is the Safe Water Network initiative in Ghana and India. Initially formed through collaborative research into water contamination, the project's evolution was driven by community co-design: researchers and local residents co-developed low-cost water filtration kiosks, changing not only infrastructure but also community perceptions of agency and health ownership (Safe Water Network, n.d.). Again, what began as an inquiry evolved into practice change—and ultimately societal transformation.

This final phase grounds the entire model. Scientific inquiry, when nurtured by 'stuporism'—experienced as a brief emotional surge of surprise that acts as an epistemic catalyst, propelling learners into deeper scientific inquiry and sustained by collaborative imagination—leads not merely to knowing and understanding, but to becoming, and importantly, to continuous becoming. It affirms that learners of all ages are not merely consumers of knowledge but co-creators of new ways of knowing and being (seeing new realities). When they are allowed to journey from awe to action, they emerge not only more informed but also more capable of reshaping the world around them. In this way, societal transformation is understood not as a downstream application of scientific knowledge, but as a cumulative outcome of altered perceptions, reconfigured practices, and redistributed agencies that emerge as inquiry moves from individual insight to collective action.

The Magic & 'Stuporism' Pedagogical Toolbox

Having outlined the **six-step framework** for 'Stuporism'-Driven Scientific Inquiry and Societal Transformation—from the initial shock when encountering a phenomenon to the enactment of material and societal change—we now turn to a rich source of practical strategies that can breathe life into this process: the art of magic and illusion. While often dismissed as mere entertainment, magic operates through deeply educational principles—disrupting perception, triggering epistemic surprise, and reorienting attention in ways that mirror the early stages of inquiry. Magicians, like great educators and scientists, are architects of cognitive dissonance and perspectival shift. Their techniques offer us more than metaphors—they offer tools.

This approach resonates with pedagogical innovations in higher education, such as those explored in the Creative Learning for Boosting Bio-Economy within HEIs' Curricula (CL4Bio) initiative. This program demonstrates how creativity, narrative-based inquiry, and experiential engagement—including elements of surprise and embodied thinking—can cultivate future-oriented mindsets, critical reflection, and sustainability literacy among students (Symeonidou, Fiore, Audiello, et al., 2023) (CL4Bio, 2025). Such initiatives show that fostering wonder and shifting perceptions are not only desirable in education but essential for equipping learners to navigate and reimagine complex socio-economic systems.

In what follows, we present a **curated toolbox of methods assembled** from magic, misdirection, and perceptual play that educators, facilitators, entrepreneurs, and/or change agents can adapt to activate 'stuporism', sustain wonder, and guide learners towards **transformative inquiry** grounded in empirical evidence (see Fig. 10.2).

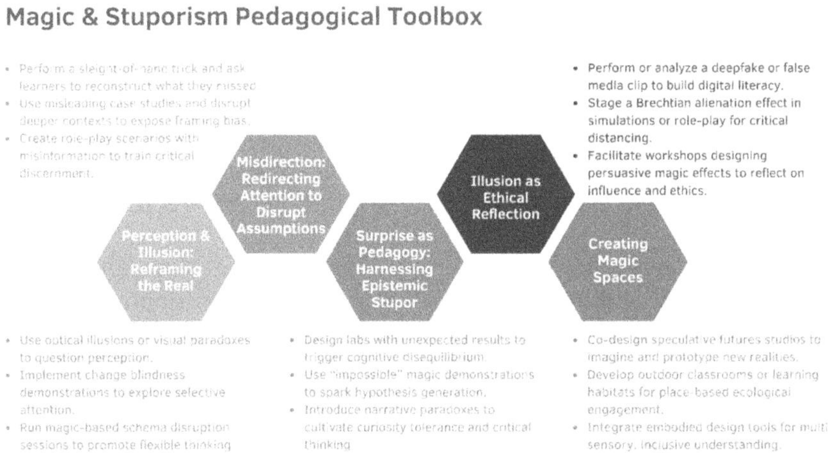

Fig. 10.2 The Magic & 'Stuporism' Pedagogical Toolbox brings together a curated set of pedagogical strategies drawn from magic, misdirection, perceptual play, and design-based inquiry. It offers educators, facilitators, and change agents a compact framework they can "carry in their pockets" and draw from when designing transformative learning experiences

Perception and Illusion: Reframing the Real

Magicians craft illusions by exploiting perceptual shortcuts—our brain fills gaps to create coherence (Kuhn et al., 2008). These interventions reveal the constructed nature of experience. A practical educational analogue is Stanford School's Designing for Social Systems workshops, where professionals use framing disruptions to uncover hidden power dynamics in social systems (Both, 2018). These disruptions lead directly to reframing possibilities—what may be rather than what is—mirroring the initial shock in our framework.

Tool A: Use Optical Illusions or Visual Paradoxes. Ambiguous images (e.g., Rubin's Vase, the Duck–Rabbit), impossible figures (e.g., Penrose triangle), vanishing objects, or distorted perspectives (e.g., Ames Room illusions).

Present one of these images without context and prompt learners with simple prompts such as: What do you see? What else could it be? What's going on here? Follow up by connecting the visual phenomenon to the idea that perception is not passive reception but active interpretation.

In science classrooms, these illusions can anchor lessons on visual cognition, data interpretation, or phenomenology. In social studies or media literacy, they can be metaphors for biased narratives, incomplete data, or alternative viewpoints.

Kuhn et al. (2008) argue that optical illusions mimic real-world conditions under which people misperceive reality—offering a hands-on entry point into understanding cognitive biases and the constructed nature of knowledge.

The outcome may be that students will become more conscious of perceptual fallibility and develop a habit of questioning first impressions and surface appearances.

Tool B: Implement Change Blindness Demonstrations. Use "spot the difference" animations or Rensink's flicker paradigm (Rensink, 2001), where two alternating images contain a subtle but significant change—yet students often fail to detect it immediately. The "door study", where people fail to notice being switched with a stranger mid-conversation, can be discussed or reenacted in video form.

Use in psychology or ethics classes to raise awareness of how selective attention operates—and fails. Students can record their initial impressions, then reflect on what was missed and why. Teachers can then scaffold into discussions about attention in high-stakes environments: e.g., decision-making in medicine, law enforcement, or public policy. Rensink and Kuhn (2015) show that change blindness reveals not only perceptual limitations but also how beliefs and expectations guide what we think we see. Such moments of perceptual failure are essential for cultivating epistemic openness and humility.

The outcome can be that students will develop heightened awareness of attentional framing, confirmation bias, and how what we notice is shaped by what we expect.

Tool C: Magic-Based Schema Disruption Session. Begin a unit with a magic trick—such as the "floating ring" or a simple "coin vanish". Ask students to individually or in groups construct explanations using what they know from science, math, or psychology. Then introduce new information that alters the explanation. Repeat the cycle.

This can be very effective in STEAM subjects, where students are required to generate hypotheses, model unseen mechanisms, and reconcile anomalies. It's also powerful in design thinking, as the experience shifts learners into a mindset of curiosity and playful inquiry. Li and McCalla (2020) developed a pedagogical model using magic tricks to support creative design education. Their results show that such interventions stimulate divergent thinking, encouraging students to consider multiple solutions and question rigid schemas. The magic serves as a liminal space between known and unknown—ideal for idea generation and inquiry.

The outcome can be that students will become more flexible in their thinking, engage in systems-level modelling, and develop a mindset of exploratory possibility, key for innovation.

Misdirection: Redirecting Attention to Reveal Assumptions

Magic's misdirection steers attention away from method to create surprise—an act that parallels biased framing in public discourse. Icard and Fervari (2024) describe misdirection as belief reconfiguration, a concept mirrored in Michigan's Problem-Solving Initiative, where students initially act on misleading case data before uncovering systemic truth (Michigan PSI, n.d.). These activities strengthen critical thinking by making explicit the process through which interpretations are constructed.

Tool A: Conduct a Sleight-of-Hand Trick, then Ask Learners to Reconstruct What They Missed. A classic vanishing object trick (e.g., the 'French drop') is performed at the start of a lesson. Students are asked: What happened? Where did your attention go? What assumptions did you make?

This tool can be used in psychology, media literacy, or even science classrooms to introduce themes such as selective attention, misinformation, and belief perseverance. Educators may record the trick from different angles or replay it slowly to reveal hidden movements, inviting learners to reflect on how and why their interpretations diverged from reality. Icard and Fervari (2024) suggest that misdirection operates through dynamic belief revision rather than simply redirecting attention. Asking students to reconstruct events enhances their epistemic vigilance—a key metacognitive skill.

The outcome may be that learners will gradually become more aware of how beliefs are formed, how easily they can be misled, and how to question initial interpretations.

Tool B: Use Misleading Case Studies with False Leads, Then Reveal the Deeper Context. Present students with a short news article or data set that contains a misleading narrative—such as a correlation mistaken for causation, or a one-sided explanation. Once they reach conclusions, introduce new context or contradictory data that reframes the case.

Application: This tool is ideal for social sciences, environmental education, law, and business ethics. For example, in a climate change module, a cherry-picked graph might downplay global warming until the full data sets are revealed, prompting students to re-examine their interpretation.

The Problem-Solving Initiative at the University of Michigan uses legal and social problem cases embedded with misleading data to help students uncover deeper systemic structures (Michigan PSI, n.d.). The pedagogical power lies not in 'correcting' students but in revealing how framing guides what we think we know (Kahneman, 2011).

The outcome of this may be that students will build resilience to manipulation, recognise framing bias, and develop critical literacy towards data and narrative.

Tool C: Create Role-Play Scenarios Where Misinformation Is Intentionally Introduced. In small groups, students participate in a mock town hall or UN negotiation. One or more participants receive deliberately biased or misleading briefing materials (e.g., slanted news reports, falsified statistics). After the simulation, the class reflects on how these distortions influenced discourse and decision-making.

Effective in politics, media, ethics, and civic education courses, it teaches not only how misinformation spreads but also how it feels to encounter and manage it in real time. The embodied nature of the experience strengthens memory and ethical reflection. Role-play combined with misinformation exposure cultivates what McGrew et al. (2018) call lateral reading: a digital literacy skill in which learners actively seek context and corroboration rather than passively absorb claims. This supports the development of narrative scepticism without inducing cynicism.

The outcome of this can be that learners will develop ethical discernment, improve their attention to narrative framing, and begin to track the influence of partial information on collective decision-making.

Surprise as Pedagogy: The Role of Epistemic Stupor (or The Epistemology of Stupor)

'Stuporism' arises when expectations collapse—when what we know fails us and the world momentarily defies our understanding. Importantly, the pedagogical efficacy of surprise does not depend on learners' explicit awareness of their own expectations or on metacognitive preparedness.

Instead, it leverages the disruption itself, treating this cognitive rupture not as an error but as the moment through which expectations are surfaced, questioned, and opened to inquiry. Like a magician's illusion, surprise disrupts our cognitive schemas, creating fertile ground for inquiry. The educational use of surprise—whether through contradiction, illusion, or paradox—can lead to richer questioning, collaborative exploration, and a deeper comfort with uncertainty.

As seen in initiatives such as the Exploratorium Teacher Institute (Exploratorium, n.d.) and Seeds to Solutions in California (TenStrands, n.d.), carefully designed moments of 'impossibility' provoke wonder and transform learners from passive recipients into active questioners and community actors. In the Seeds to Solutions programme, middle-school students explored environmental issues by designing solutions for their communities, such as rain gardens or local air-quality monitors. These interventions are driven by epistemic surprise: students learn that their environment is not as stable or healthy as assumed! This sense of wonder and shock reframes their roles, encouraging them to move from spectators to agents of change (TenStrands, n.d.). This surprise fosters ownership of learning, as students realise they can actively participate in shaping their communities through evidence-based action. Below, we elaborate on tools that harness surprise for transformative learning.

Tool A: Design Labs with Unexpected Results. Present students with an experiment that contradicts intuition—e.g., placing a heavy steel ball on a bed of nails or using supercooled water that remains liquid until disturbed. Alternatively, reverse a common procedure and ask students to predict the outcome, only to witness something unexpected.

This tool can be used in physics, chemistry, and biology to introduce concepts such as surface tension, fluid dynamics, or chemical reactions. In the humanities, similar surprise setups can be created by introducing a 'twist' into a historical narrative or philosophical argument. The Exploratorium's pedagogy emphasise surprise as a means of generating cognitive disequilibrium, a moment when learners' existing mental models are no longer sufficient. This disequilibrium prompts productive struggle, which Vygotsky and Cole (1978) and Piaget and Cook (1952) identify as key to conceptual change.

The outcome of this may be that students will shift from passive consumption to active theorising, engaging more deeply with material through surprise-triggered robust inquiry.

Tool B: Use 'Impossible' Magic Demonstrations to Spark Hypothesis Generation. Perform a simple illusion (e.g., levitation of a small object using magnets and hidden supports, or a water glass turned upside-down without spilling). Invite learners to propose multiple scientific explanations, rank them, and test the feasibility of each.

This tool can be used in early science education, maker spaces, and critical thinking workshops. It is ideal for bridging intuitive beliefs and scientific reasoning, particularly when students initially accept 'magical' explanations before constructing testable hypotheses. Symeonidou & Audiello describe the use of illusions in science education as an effective way to sustain student engagement and foster epistemic curiosity (The Magic Academy n.d.). The trick destabilises prior knowledge, while the follow-up task invites students to construct alternative explanations through the lens of scientific inquiry.

The outcome can be that learners develop a process-oriented mindset, where wonder leads to structured exploration rather than passive amazement.

Tool C: Use Narrative Paradoxes. Present paradoxes such as "This sentence is false", or Zeno's paradoxes of motion, and invite students to discuss their implications in philosophy, logic, or even real-world policymaking. You can also introduce moral paradoxes (e.g., the trolley problem) and artistic contradictions (e.g., M.C. Escher's staircase drawings).

This tool can be used in philosophy, ethics, literature, and logic courses, as well as in science or maths to highlight epistemological uncertainty. These activities foster curiosity tolerance—the ability to stay with a complex, unresolved idea. Research in gifted and philosophical education (Lipman, 2003; Danek & Wiley, 2017) shows that engagement with paradoxes promotes metacognitive reflection, creative reasoning, and openness to ambiguity—traits essential for critical and scientific thinking.

The outcome can be that learners will become more comfortable with uncertainty, deepen dialogic reasoning, and experience the productive tension of unresolved inquiry.

Illusion as Ethical Reflection

Illusion is not only a tool for surprise—it is a structure of influence. In politics, advertising, and digital media, illusion and misdirection are regularly employed to shape public sentiment, obscure power dynamics, and reinforce dominant narratives. 'Stuporism', in this context, becomes not merely a moment of wonder but an awakening to the mechanisms of manipulation.

By borrowing techniques from Brechtian theatre, magic, and media analysis, educators can guide students towards meta-awareness of how perception is shaped, disrupted, and ethically leveraged.

This critical stance is closely aligned with transdisciplinary pedagogy, where the arts and sciences are not isolated in separate silos but taught together to foster aesthetic and epistemic openness. As Burnard, Colucci-Gray, and Sinha (2021) argue, the arts illuminate and enrich our lives and deepen our understanding of who we are and how we make sense of the world (see also Burnard et al., Chap. 5, this volume). Hence, teaching the arts and sciences together enables learners to explore how knowledge systems and transversal skills that apply across domains are constructed, communicated, and challenged. Illusion, in this light, becomes not a deception but a lens—a pedagogical technique that helps illuminate the architecture of belief and the ethics of persuasion.

Tool A: Perform or Analyze a Deepfake or False Media Clip. Present a short, convincing deepfake or fabricated news story (created or borrowed from verified datasets like MIT's 'Detect Fakes' or 'Media Literacy Lab' resources). Ask students to respond emotionally and cognitively to the clip. Then, reveal its inauthenticity and analyze the techniques used to persuade or manipulate.

This tool can be used in media studies, digital ethics, or political science classrooms, as it fosters digital literacy and awareness of how persuasion, trust, and authenticity are constructed in the post-truth era. Research by Wardle & Derakhshan (2017) on 'information disorder' highlights how the emotional appeal and visual plausibility of fake media override critical judgement. When students are first 'tricked' by a clip and then walk back through their reasoning, they become better equipped to detect and resist misinformation.

The outcome may be that students will become more sceptical towards surface appearances, understand rhetorical framing, and develop cognitive resilience in a disinformation-saturated media ecosystem.

Tool B: Stage a Brechtian 'Alienation Effect' in Simulations or Role Play. During a class role-play (e.g., a courtroom, town hall, or historical debate), have participants break character mid-activity to critically reflect on the scene—its realism, its ideological framing, and whose voices are missing. For example: "Who is being silenced here?" or "What assumptions are we reinforcing?"

This tool can be ideal for history, drama, politics, or cultural studies. In educational terms, this effect of audiences becoming engaged in reflection cultivates epistemic distance, a stance from which learners critically evaluate the systems and stories they are part of (Boal, 2000; Schutzman & Cohen-Cruz, 1994).

The outcome will be that learners will develop the ability to step outside scripted ideologies, question constructed narratives, and become critically literate citizens capable of recognising when power is disguised as entertainment or neutrality. A notable example is the 'Narratives of Crisis' course, developed at the London School of Economics. This interdisciplinary course deconstructs media narratives about economic collapse, political instability, and public trauma. Using techniques from literary studies, media theory, and even theatrical illusion, students investigate how crisis is constructed—not just experienced. Like a magic trick, the crisis becomes visible only when you see what the magician wants you to miss: underlying structures, systemic neglect, or historical patterns.

The outcome can be that students will become adept at narrative deconstruction, reframing what they're told into a critical understanding of why they're being told it.

Tool C: Facilitate a Workshop where Students Design and Deconstruct a Persuasive Magic Effect.

Students design a simple magic trick (e.g., a force card selection or a prediction trick) intended to influence a peer's decision. After performing it, they reflect on the strategies used—gestural misdirection, linguistic priming, attention control—and compare them with techniques in advertising, speeches, or politics.

This tool can be used in rhetoric, creative communication, leadership training, or even business ethics courses. It combines practical creativity with ethical reflection. Kuhn et al. (2008) argue that magic exposes the mechanisms of belief formation and decision-making. When students become 'architects of illusion', they gain a deeper understanding of the ethical boundaries of persuasion and the malleability of perception.

The outcome may be that students not only become more skilled at identifying manipulation but also engage in ethical self-reflection, creating an inner voice that asks "When is influence acceptable, and when does it become coercion?"

Creating Magic Spaces: Pedagogical Design for Possibility

In magic, the stage is never incidental. Every angle, pause, silence, and spotlight is part of the illusion—carefully crafted to heighten anticipation, direct attention, and create space for astonishment. The magician doesn't merely perform within a setting; they design a world in which the impossible feels momentarily real.

Education, too, has such potential. The design of learning spaces—physical, conceptual, and emotional—is a form of invitation. It can shape learners not only into curious observers, but into possibility thinkers: individuals who imagine beyond the present constraints, ask "what if?" and "what else?" and begin to explore futures not yet formed.

As Burnard (2006) describes in her work on possibility thinking, this shift emerges when learners move from "what is" to "what might be", often through playful questioning, improvisation, and experimentation. In all cases, we come to see that another possible is possible. When educators draw on literatures in science and technology, not as bodies of knowledge but as material artefacts *in the world*, learners are invited to pay attention to the materiality of magic spaces, the affordance of sensorial perceptions and the lived experience of possibility. Together, as learning communities, they deliberately create 'magic spaces', they do more than provoke curiosity; they foster a mindset where transformation—personal, social, or scientific—feels not only plausible but imminent.

These spaces are thresholds, staging grounds for 'stuporism'—fueled inquiry, where perception bends, norms loosen, and students can reimagine both their roles and their realities. What follows are three pedagogical strategies that turn such spaces into platforms for enacting future possibility.

Tool A: Co-design a Speculative Futures Studio. Students create 'future labs' or 'possibility studios' where they explore big societal questions—such as the future of energy, migration, or AI. They use design fiction, speculative maps, and world-building to imagine preferred futures, then reverse-engineer paths to get there.

This tool is well suited to interdisciplinary and transdisciplinary curricula that include civic education, sustainability, design, and digital literacies (McGrew et al., 2017). Educators facilitate sessions in which students prototype potential systems rather than merely analysing existing ones. Consider 3D-printed city models, alternate-history timelines, or interactive ethics simulations. The Seeds of Change project empowered students to move from awareness to material invention—transforming schoolyards into green infrastructure through inquiry-driven design (European School Education Platform, n.d.). Similarly, Ratti & Claudel (2016) emphasise speculative design as a pathway to civic imagination, where learners move from "what is" to "what if".

The outcome may be that students will learn affordance articulation, how to perceive possibilities embedded in their environment, and develop futures literacy which is the capacity to imagine, assess, and influence potential futures.

Tool B: Develop an Outdoor Classroom or Learning Habitat. Students are invited to co-create an outdoor space—building structures like natural shelters, pollinator gardens, rainwater systems, or artistic installations that reflect ecological principles. Learning is embedded in the act of place-making.

This tool can be applied across science, ecology, civic engagement, and art education. The process becomes transdisciplinary: maths is used to measure rainfall catchment; literature reflects on nature writing; science analyses biodiversity patterns; and civic skills emerge in planning permissions or budgeting. The Boston Schoolyard Initiative (BSI, 2020) is a benchmark programme in which underutilised schoolyards were transformed into learning laboratories through participatory design. These spaces democratized knowledge and returned students to the world as active environmental stewards, not just spectators.

The outcome can be that place-based learning, ecological awareness, and agency in the built environment will be enhanced. Students will see themselves not only as learners but also as co-designers of their futures.

Tool C: Integrate Embodied Design Tools for Abstract Thinking. Use physical manipulatives, such as foam blocks to simulate statistics, rope to model network theory, or full-body simulations (e.g., acting out algorithms or protein synthesis). Pair these with reflective sketching or digital modelling to support multimodal cognition.

This approach can be particularly effective in maths, science, and systems thinking. For example, in an 'embodied data' session, students form bar charts with their bodies, or in philosophy class, use spatial movements to model ethical dilemmas. The Embodied Design Research Laboratory (EDRL, 2025) argues that hands-on interaction with abstract concepts strengthens retention, reasoning, and the sensorimotor grounding of complex ideas (Abrahamson & Lindgren, 2014). These tools invite learners into a dialogue with space and self, turning learning into an aesthetic and kinesthetic experience.

The outcome may be improved conceptual understanding through multisensory engagement, support for inclusive learning styles, and greater transfer of learning across domains.

We present this 'magic toolbox' with the hope that educators, facilitators, and change agents will use this compact framework they can "carry in their pockets" and draw inspiration from when designing transformative learning experiences. Each cluster highlights practical tools that activate stupor, sustain wonder, and guide learners towards curiosity-driven inquiry, critical reflection, and imaginative possibility-thinking. Designed for flexible use across disciplines, the toolbox can be put to work in classrooms, workshops, and community contexts to provoke new ways of seeing, challenge assumptions, and turn knowledge into action—helping learners become co-creators of change rather than passive recipients of information.

The 'Vitality of Stuporism' as an Epistemic Catalyst

The proposed model offers a framework for designing transformative learning environments that cultivate social change. By embedding the awe of illusion, the aesthetics of art, and the ethics of inquiry into this stepwise model, we restore to education what too often gets lost: wonder with consequence.

Why? Because sometimes the most powerful way to change the world is to be astonished by it first! This is the vitality of 'stuporism'! This is the vitality of the momentary surge of surprise that catalyses deeper scientific inquiry. This is the vitality that enacts a pedagogy not of certainty, but of openings; an invitation to dwell in the in-between, where science becomes wonder and wonder becomes action.

We offer this framework as a provocation and a promise: from Pam, who crafts spaces where arts and science creativities become transversal inquiry; from Antonia, who engineers pathways for science to speak to minds; and from Danilo, who performs the line between illusion and wonder.

Together, we ask not only how change happens, but how it begins—in awe, in inquiry, and in the refusal to stop asking what else the world could become. Because the most powerful way to enact societal change may be to begin by preserving the questions we ask ourselves.

References

Abrahamson, D., & Lindgren, R. (2014). Embodiment and embodied design. In R. K. Sawyer (Ed.), *The Cambridge handbook of the learning sciences* (2nd ed., pp. 358–376). Cambridge University Press.

Audiello, D., & Symeonidou, A. (2024). Stuporism: The philosophy of wonder. In *Encyclopedia of quality of life and well-being research* (pp. 6923–6926). Springer International Publishing.

Blackiston, D. J., Silva Casey, E., & Weiss, M. R. (2008). Retention of memory through metamorphosis: Can a moth remember what it learned as a caterpillar? *PLoS One, 3*(3), e1736.

Boal, A. (2000). *Theater of the oppressed*. Pluto Press.

Bonney, R., Cooper, C. B., Dickinson, J., Kelling, S., Phillips, T., Rosenberg, K. V., & Shirk, J. (2009). Citizen science: A developing tool for expanding science knowledge and scientific literacy. *BioScience, 59*(11), 977–984.

Boston Schoolyard Initiative. (2020). Designing outdoor classrooms for equity and sustainability. https://schoolyards.org

Both, T. (2018, March 9). *Human-centered, systems-minded design*. Stanford Social Innovation Review. http://www.ssir.org/articles/entry/human_centered_systems_minded_design

Brecht, B. (1964). *Brecht on theatre: The development of an aesthetic* (J. Willett, Ed. & Trans.). Hill and Wang.

Burnard, P., Colucci-Gray, L., & Sinha, P. (2021). Transdisciplinarity: Letting arts and science teach together. *Curriculum Perspectives, 41*(1), 113–118.

Burnard, P., Craft, A., Cremin, T., Duffy, B., Hanson, R., Keene, J., Haynes, L., & Burns, D. (2006). Documenting 'possibility thinking': A journey of collaborative enquiry. *International Journal of Early Years Education, 14*(3), 243–262.

CL4Bio. (2025). *Creative learning for boosting bio-economy within HEIs' curricula*. https://cl4bio.web.ua.pt

Dadachova, E., Bryan, R. A., Huang, X., Moadel, T., Schweitzer, A. D., Aisen, P., Nosanchuk, J. D., & Casadevall, A. (2007). Ionizing radiation changes the electronic properties of melanin and enhances the growth of melanized fungi. *PLoS One, 2*(5), e457. https://doi.org/10.1371/journal.pone.0000457

Danek, A. H., & Wiley, J. (2017). What about false insights? Deconstructing the aha! Experience along its multiple dimensions for correct and incorrect solutions separately. *Frontiers in Psychology, 7*, 2077.

Desmond, W. (2016). *The intimate universal: The hidden porosity among religion, art, philosophy, and politics*. Columbia University Press.

Embodied Design Research Lab (EDRL). (2025). https://edrl.berkeley.edu.adri-ivs.com

European School Education Platform. (n.d.). Seeds of change. *eTwinning*. https://school-education.ec.europa.eu/en/etwinning/projects/seeds-change

Exploratorium. (n.d.). Teacher institute. *Exploratorium*. https://www.exploratorium.edu/education/teacher-institute
Freire, P. (1994). *Pedagogy of hope: Reliving pedagogy of the oppressed*. Continuum.
Gibson, J. J. (1979). *The ecological approach to visual perception*. Psychology Press.
Glasser, O. (1993). *Wilhelm Conrad Röntgen and the early history of the roentgen rays* (Vol. No. 1).
Gordon, D. M. (2010). *Ant encounters: Interaction networks and colony behavior*. Princeton University Press.
Greene, B. (2004). *The fabric of the cosmos: Space, time, and the texture of reality*. Knopf.
Heidegger, M. (1962). *Being and time* (J. Macquarrie & E. Robinson, Trans.). Blackwell. (Original work published 1927).
Hmelo-Silver, C. E., Duncan, R. G., & Chinn, C. A. (2007). Scaffolding and achievement in problem-based and inquiry learning: A response to Kirschner, Sweller, and Clark (2006). *Educational Psychologist, 42*(2), 99–107.
Icard, B., & Fervari, R. (2024). Beyond the spell: A dynamic logic analysis of misdirection. *arXiv preprint* arXiv:2401.14516.
Jinek, M., Chylinski, K., Fonfara, I., Hauer, M., Doudna, J. A., & Charpentier, E. (2012). A programmable dual-RNA–guided DNA endonuclease in adaptive bacterial immunity. *Science, 337*(6096), 816–821.
Kahneman, D. (2011). *Thinking, fast and slow*. Farrar, Straus and Giroux.
Keltner, D., & Haidt, J. (2003). Approaching awe, a moral, spiritual, and aesthetic emotion. *Cognition and Emotion, 17*(2), 297–314.
Krasny, M. E., & Tidball, K. G. (2012). Civic ecology: A pathway for earth stewardship in cities. *Frontiers in Ecology and the Environment, 10*(5), 267–273.
Kuhn, G., Amlani, A. A., & Rensink, R. A. (2008). Towards a science of magic. *Trends in Cognitive Sciences, 12*(9), 349–354.
Li, T., & McCalla, L. E. (2020). The design and development of a magic-based teaching method in facilitating creative design thinking. *International Journal of Designs for Learning, 11*(3), 67–77.
Lintott, C. J., Schawinski, K., Slosar, A., Land, K., Bamford, S., Thomas, D., Raddick, M. J., Nichol, R. C., Szalay, A., Andreescu, D., Murray, P., & Vandenberg, J. (2008). Galaxy zoo: Morphologies derived from visual inspection of galaxies from the Sloan digital sky survey. *Monthly Notices of the Royal Astronomical Society, 389*(3), 1179–1189.
Lipman, M. (2003). *Thinking in education*. Cambridge University Press.
Loewenstein, G. (1994). The psychology of curiosity: A review and reinterpretation. *Psychological Bulletin, 116*(1), 75.
McGrew, S., Breakstone, J., Ortega, T., Smith, M., & Wineburg, S. (2018). Can students evaluate online sources? Learning from assessments of civic online reasoning. *Theory & Research in Social Education, 46*(2), 165–193.

McGrew, S., Ortega, T., Breakstone, J., & Wineburg, S. (2017). The challenge that's bigger than fake news: Teaching students to engage in civic online reasoning. *American Educator, 41*(3), 4.

Mezirow, J. (2015). *Transformative learning* (p. 319). Challenging Educational Theories.

Michigan Problem Solving Initiative. (n.d.). *The problem solving initiative: Teaching through complex legal and policy problems.* University of Michigan Law School. https://michigan.law.umich.edu/academics/centers-and-programs/problem-solving-initiative-psi

Moberg, F., & Folke, C. (1999). Ecological goods and services of coral reef ecosystems. *Ecological Economics, 29*(2), 215–233. https://doi.org/10.1016/S0921-8009(99)00009-9

Opdal, P. M. (2001). Curiosity, wonder and education seen as perspective development. *Studies in Philosophy and Education, 20*(4), 331–344.

Piaget, J., & Cook, M. (1952). *The origins of intelligence in children* (Vol. 8, No. 5, pp. 18–1952). International Universities Press.

Ratti, C., & Claudel, M. (2016). *The city of tomorrow: Sensors, networks, hackers, and the future of urban life.* Yale University Press.

Rensink, R. A. (2001). Change blindness: Implications for the nature of visual attention. In *Vision and attention* (pp. 169–188). Springer New York.

Rothstein, D., & Santana, L. (2011). *Make just one change: Teach students to ask their own questions.* Harvard Education Press.

Safe Water Network. (n.d.). *Safe water network.* https://safewaternetwork.org/

Sánchez, C. (2020). Fungal potential for the degradation of petroleum-based polymers: An overview of macro-and microplastics biodegradation. *Biotechnology Advances, 40*, 107501.

Schatzki, T. R. (2019). *Social change in a material world.* Routledge.

Schinkel, A. (2017). The educational importance of deep wonder. *Journal of Philosophy of Education, 51*(2), 538–553.

Schutzman, M., & Cohen-Cruz, J. (1994). *Playing boal: Theatre, therapy, activism.* Routledge.

Sheldrake, M. (2021). *Entangled life: How fungi make our worlds, change our minds & shape our futures.* Random House Trade Paperbacks.

Simard, S. W., Perry, D. A., Jones, M. D., Myrold, D. D., Durall, D. M., & Molina, R. (1997). Net transfer of carbon between ectomycorrhizal tree species in the field. *Nature, 388*(6642), 579–582.

Symeonidou, A., Abbas, M., & Audiello, D. (2025). Chapter 8 science communication: Re-engineering the perception of (Im)possibilities in teaching and learning. In *Eruptive research.* Brill.

Symeonidou, A., Fiore, M., Colantuono, F., Audiello, D., & Burnard, P. (2023). Creative problem-solving practices amplifying threshold concepts of 'green','circular'and 'bio economy': Emerging evidence from an instrumental case study of modules involving undergraduate Italian economics students. In *Book of abstracts of open workshop in 'sharing good practices in the field of creative learning and bio-economy from invited and partners' organizations'*. UA Editora University of Aveiro.

Taylor, R. P., Micolich, A. P., & Jonas, D. (1999). Fractal analysis of Pollock's drip paintings. *Nature, 399*(6735), 422–422.

Ten Strands. (n.d.). https://tenstrands.org/

The Magic Academy. (n.d.). Magic academy: Teaching life skills through magic. https://www.magicacademy.co.uk/

Verger, N. B. (2025). The aesthetics of the possible. *Possibility Studies & Society*, 27538699251318068.

Verger, N. B., Duymedjian, R., Wegener, C., & Glăveanu, V. P. (2024). Creative preservation: A framework of creativity in support of degrowth. *Review of General Psychology, 28*(3), 268–281.

Vygotsky, L. S., & Cole, M. (1978). *Mind in society: Development of higher psychological processes.* Harvard University Press.

Wardle, C., & Derakhshan, H. (2017). *Information disorder: Toward an interdisciplinary framework for research and policymaking* (Vol. 27, pp. 1–107). Council of Europe.

CHAPTER 11

Wonder, Education, and Possible Futures

David Mattson

Introduction

> To consider oneself a wonder and reflect upon one's uniqueness has been described as "humility in the face of life, because life represents the rarest and most astonishing creations observable in the universe." Others too are a wonder, not a mystery to be solved or an alien to be dissected, analyzed, or totalized, but a wonder that exists to be appreciated, discovered, and rediscovered, puzzled over, even doubted, but never to be known completely and fully, so that a future potential, possibility, or story is encouraged and not denied. (Kearns, 2015, p. 99)

The centrality of wonder within education, far from a new idea, is often traced to the teachings of Socrates. The infamous quote "philosophy begins in wonder" reverberates through the writings of philosophers no less influential than Descartes, Spinoza, Arendt, and Dewey (Lloyd, 2018). Philosophically, wonder broadens our awareness of what we know and the limits of our knowledge (Lacey, 2016). Such experiences facilitate our

D. Mattson (✉)
Freddie Reisman Center for Translational Research in Creativity and Motivation, Drexel University, Philadelphia, PA, USA
e-mail: dmm563@drexel.edu

N. B. Verger, D. Mattson (eds.), *Possibility Studies and Societal Futures*, Palgrave Advances in Possibility Studies,
https://doi.org/10.1007/978-3-032-26379-7_11

critical thinking and meaning-making processes, and thus have long been considered central to the purpose of education (Dewey, 1934; d'Agenes, 2019). Beyond a pedagogical resource for learning, wonder informs our conceptualisation of what is and ought to be. Wonder arising from the acute awareness of moral hypocrisy leads us to question our social priorities and cultural norms (Egan et al., 2013). Jan Pedersen (2020, p. 56) argues that, "Wonder plays an important part in human flourishing because it takes wonder to discover how we might flourish". Wonder, understood in this context, exists as a mechanism by which we evaluate possible futures, not merely the future of our own life, but the possible futures for our society (Glăveanu, 2020).

Wonder, Imagination, and Social Change

Wonder plays a unique role in shaping societal futures by serving as the catalyst for questioning the presuppositions of our cultural norms and rules (Opdal, 2001; Prinz, 2013). Experiences of wonder introduce the mind to new perspectives and draw us into a conversation with the possible. As Arendt (1978) and Lloyd (2018) argue, wonder is necessary for questioning political ideology and recognising the inadequacy of privileged certainties within our society. Glăveanu (2020, p. 152) sums up the importance of wonder for shaping society, saying that wondering is "indispensable for a healthy social and political life because it makes people really engage, perhaps for the first time, with the world of possibilities surrounding them". Here, Glăveanu (2020) expresses the interdependence of wonder as a social/individual/cultural phenomenon. Wonder begets this interconnectedness, as wondering about how one might or ought to live is inseparable from social and cultural considerations (Lloyd, 2018; Peterson & Seligman, 2004). Thus, wonder, as relates to possible societal futures, serves as the catalyst for imagination and critical thinking, ideally developed during the formative years of our education.

Wonder and Education

Unsurprisingly, education, as an incubator of critical thinking, has a storied connection with wonder (Schinkel, 2017) as the Socratic approach within education draws upon the rich history of wonder in scholastic antiquity (Di Paolantonio, 2018). Researchers have identified the benefits of wonder in education to include aiding in meaning-making and increasing "students' intrinsic motivation to learn and to investigate, as well as their general interest in the world" (Conijn et al., 2022, p. 424). Robert Fuller describes the role of wonder in thinking as follows,

> As such, wonder is the ontogenetic basis for the development of thinking that goes beyond the constraints of the immediate physical environment. It stimulates our efforts to contemplate ever-more general orders of existence. The motivational functions performed by wonder, then, are closely linked with the development of some of humanity's highest cognitive and cultural achievements. Not the least of these achievements is the emergence of humanity's capacities for both moral reflection and philosophical speculation. (p. 70)

Fuller's description elevates the importance of wonder for education and the development of societal futures by linking the motivational influence of wonder with past intellectual, and social achievements. Ultimately, wonder possesses a unique capacity for encouraging students to "suspend judgment and set aside existing—potentially limiting—ways of thinking, seeing, and categorizing" (Schinkel, 2017, p. 538). Despite the many benefits of wonder for students and the historical connection between wonder and education, the phenomenon of wonder remained largely underexplored among classroom-based education research until recently (Mattson, 2023). Perhaps this is because wonder has been historically difficult to define: "pinning it down to one simple conception or definition has been a confounding issue facing philosophers since the days of Aristotle and Plato" (Gilbert & Byers, 2017, p. 913).

The Trouble with Wonder

The phenomenon of wonder has been a difficult subject for education researchers to investigate due to its complexity as an ill-defined phenomenon (Mattson, 2023). The term "Wonder" operates as a recursive contradiction, meaning the concept itself performs what it describes. The very attempt to clearly identify the requisite attributes and components of "wonder" draws the researcher into an experience of philosophical wonder (Vasalou, 2015), as this familiar experience suddenly becomes strange, increasingly meaningful, yet elusive. Wonder has often been mistaken for (or used interchangeably) with curiosity and awe, and many researchers have devoted considerable effort to disambiguating these terms. While it is beyond the scope of this chapter to re-litigate those distinctions but see Symeonidou et al., Chap. 10, this volume), the following quote from Geller et al. (2020) provides a concise summary of the differences:

> Wonder is a feeling of radical appreciation for a triggering event, typically accompanied by a behavior of pausing to reflect and a motivation to reorient one's self-understanding and sense of the world (Evans, 2012; Heschel, 1951; Moore, 2005; Parsons, 1969; Prinz, 2013; Seligman, 2004; Shiota et al., 2007). By contrast with curiosity, which is understood as "a confident and focused interest to find something out," someone experiencing wonder "is struck by the strangeness or peculiarity of the things met" (Opdal, 2001, p. 331). By contrast with awe, a feeling typically elicited by information-rich stimuli and associated with a sense of the smallness of the self (Shiota et al., 2007), wonder may be elicited by an everyday event seen in a novel light, and it does not typically result in feeling physically or mentally "smaller" (Evans, 2012).

Although distinct from wonder, curiosity and awe often appear before, after, or within experiences of wonder. Glăveanu (2020, p. 176) describes this interaction, saying, "Wondering at or wondering deeply often turns into awe … wondering about, actively, can become mere curiosity … to wonder-fully requires the interrelation between these different aspects". According to Hadzigeorgiou (2011, p. 987), "It does not take one long to realize that wonder, as a state of mind, can be associated with mystery, awe, perplexity, astonishment, surprise, amazement, admiration and bewilderment. Yet the experience of a sense of wonder cannot be reduced to the experience of any one of the aforementioned elements". The complexity of distinguishing between awe, curiosity, and wonder and the lack of a broadly accepted definition of wonder are partially responsible for the lack of educational research about how wonder appears and functions within classroom learning (Conijn et al., 2022). Another obstacle to wonder research in education has been the development of standardised testing as a measure of quality education, which threatens the position of wonder as central to the purpose of education.

Wonder and Education

Many of the same researchers who have led the resurgence of research into wonder over the past two decades have also espoused strong opinions about the current education policy regarding the prioritisation of wonder in the classroom. Glăveanu (2020, p. 171) claims, "It is undeniable that current pedagogies focused on standardized assessment and, as such, on getting the one right or correct answer are antithetical to wondering". Similarly, Gilbert (2013, p. 9) describes the current policies saying "This need to wonder has been suppressed by a system of schooling that desires to maintain order at all costs even if those structures limit student joy and enthusiasm of content related material". Schinkel (2017, p. 540) comments on barriers to wonder in education saying, "There are some reasons to think that current education systems in the Western world constitute a particularly hostile environment for the experience of wonder". Moreover, Vasco d'Agnese (2019) argued that the current culture of testing and measurement threatens the role of wonder in education and challenges our social constructions about the purpose of education. However well-intentioned, the testing and accountability culture of modern Western education has seemingly undermined the historically significant role of wonder in education.

In light of the systemic obstacles to wonder in the classroom, the considerable benefits of wonder for students, and the inherent challenge that researching a phenomenon like wonder presents, I conducted a literature review to address the following emergent questions:

1. How do education researchers investigate wonder in the classroom (what types of studies and data collection methods)?
2. How do education researchers define or conceptualise wonder in the classroom?

This literature review explores the various definitions, conceptualisations, findings, and research methods of wonder in current education studies through a descriptive analysis of 26 peer-reviewed articles. The findings address the research questions via a categorisation and rich description of the articles, followed by a discussion of the definitions and conceptualisations of wonder represented in the literature.

Literature Review Design

While wonder has been philosophised and promoted in academic journals such as *The Philosophy of Education*, wonder is often considered more of a theoretical concept and therefore remains an under-investigated phenomenon in education classroom research (Mattson, 2023). This descriptive literature review (Yang & Tate, 2012) provides researchers with a current representation of how education researchers explore wonder. Researchers often use a scoping literature review for similar purposes. However, I deemed a descriptive literature review more appropriate for two reasons. First, the study's research questions require not only identification of existing research but also deep characterisation and synthesis of how wonder is conceptualised and operationalised across studies. While scoping reviews excel at mapping research landscapes and identifying gaps, they typically do not synthesise findings with the theoretical emphasis applied here (Arksey & O'Malley, 2005). This study required interpretive synthesis to make sense of disparate definitions, theoretical frameworks, and methodological approaches used to study wonder. Second, the emergent nature of wonder research in education classrooms presents unique challenges. The limited body of empirical classroom research (only 26 studies meeting inclusion criteria) required a flexible analytical approach capable of accommodating highly varied study designs (from narrative observations to quasi-experimental interventions). Since the categorisation framework developed for this review serves both descriptive and analytical purposes, allowing for thematic organisation while maintaining the nuance of individual studies' contributions, a descriptive literature review methodology presented a better fit for this study.

Article Selection and Inclusion Criteria

Searches were conducted on academic databases through EBSCO search, including ERIC, PsycINFO, PsycTests, PsycJournals, and Teacher Research database, using the keyword "Wonder". From an initial bucket of over 10,000 articles, I refined the search results to the studies within education, removing studies that mention wonder only in the title, removing studies that only use wonder as analogous to consideration or thinking (as in, "I wondered what the respondents might say"), and removing studies that did not intentionally discuss wonder as a subject or finding of the article. To warrant inclusion, articles must have addressed wonder as a phenomenon related to classroom learning and not just use the word "wonder" in a peripheral description. Additionally, and most impactfully, the inclusion criteria stipulate that all studies include the gathering or reporting of classroom data about wonder (observational, qualitative, quantitative, narrative, or otherwise).

Articles without evidence derived from researchers' direct experience (i.e., lacking direct reporting of wonder occurrences in the classroom) were not included. This criterion resulted in the exclusion of many articles theorising that wonder is central to the purpose of education; articles about wonder relying on general classroom statistics not directly gathered by the researcher; articles about wonder in relation to educational standards; and other articles solely about wonder as a generally beneficial concept within education. The majority of articles excluded from this study fall into this category of theoretical research, lacking any classroom observation or empirical data about the phenomenon of wonder.

The researcher considered adding a publication date restriction to the inclusion criteria. However, only one of the 26 selected articles was published before 1999. Inspection of the older article, published in 1988, revealed its relevance and conceptual fit with the study's design and purpose. After refining the search results by these criteria, I began analysis of the articles that explored or investigated wonder in the context of classroom learning.

Findings

Four Categories of Classroom Research about Wonder

A variety of research designs and theoretical perspectives about wonder were found among the final 26 articles. Researchers use various definitions of wonder, associated wonder with several adjacent phenomena (such as curiosity, awe, and enchantment), and explore wonder using different research designs. To assist with the descriptive literature objective of describing the various methods researchers use to investigate wonder in the classroom, the articles were divided into four categories: Observational research based on personal or collective classroom observations, Instrumental research with the explicit purpose of developing tools for identification and measurement of wonder, Participatory research reporting on lesson plans or classroom interventions that elicit wonder, and Quasi-Experimental research (includes a control group) conducted with the intention of eliciting and measuring wonder in the classroom. Table 11.1 provides a list of each article according to their respective category. Each category of article is discussed in detail with examples in the following pages. Fig. 11.1 provides a visual of the breakdown of articles by category.

Table 11.1 Four categories of wonder in the classroom literature

Quasi-experimental studies of wonder
The Garden Wonder Wall: Fostering Wonder and Curiosity on Multi-Day Garden Field Trips—Driscoll and Lownds (2007)
Using Moments of Wonder in Science with Pre-Service Teachers—Cooper and Gilbert (2016)
Wonder as a tool to engage preservice elementary teachers in science learning and teaching—Gilbert and Byers (2017)
W(h)ither the sense of wonder of pre-service primary teachers when teaching science?: A preliminary study of their personal experiences—Stolberg (2008)
The Role of Wonder in Students' Conception of and Learning About Evolution—Sundberg and Andersson (2023)
Fostering a Sense of Wonder in the Science Classroom—Hadzigeorgiou (2012)
Teaching about the importance of trees: A study with young children—Hadzigeorgiou (2011)
Participatory studies of wonder
A Wonder-Full Task Leads to a Wonder-Full Intervention—Ruttenberg-Rozen (2020)
An Expanded Observe-Wonder-Learn: Moving Observe-Wonder-Learn from Assessing Prior Knowledge into a Unit-Planning Tool—Baird and Coy (2020)
Implementing an Imaginative Unit: Wonders of the Water Cycle—Hrennikoff (2006)
Methods & Strategies: I Wonder ... —Stevenson (2013)
Sense of Wonder Science—Potter and Ritz (2006)
Teaching secondary mathematics lessons for joy and wonder—Scheitlin et al. (2023)
The magic of wondering: Building understanding through online inquiry—Coiro (2015)
Try This: Rock Wonder—Preston (2019)
Working Wonders—Bianchi (2013)
Observational studies of wonder
The Role of Teacher Stance When Infusing Inquiry Questioning Into Middle School Science Classrooms—Mackenzie (2001)
Building a Sense of Wonder in Secondary Content Literacy Instruction—Wold and Elish-Piper (2008)
Unconventional First Days: Encouraging Students To Wonder about Social Life and Learning—Higgins (1999)
Bringing the Jellyfish Home: Environmental Consciousness and "Sense of Wonder" in Young Children's Encounters with Natural Landscapes and Places—Jørgensen (2016)
Experiencing the Sacred in Dance Education: Wonder, Compassion, Wisdom, and Wholeness in the Classroom—Debenham & Debenham (2008)
Making the Science Classroom a Place for Wonder—Niedermeyer (2018)
Where is Wonder?—Burton (2013)
Instrumental studies of wonder
Fostering Wonder in Young Children: Baseline Study of Two First Grade Classrooms –McWilliams (1999)
Development and Validation of a Capacity for Wonder Scale for Use in Educational Settings—Geller et al. (2020)
A theoretical framework and questionnaire for wonder-full Education—Conijn et al. (2022)

Note. This table organises the literature on wonder in classroom settings into four methodological categories based on research approach

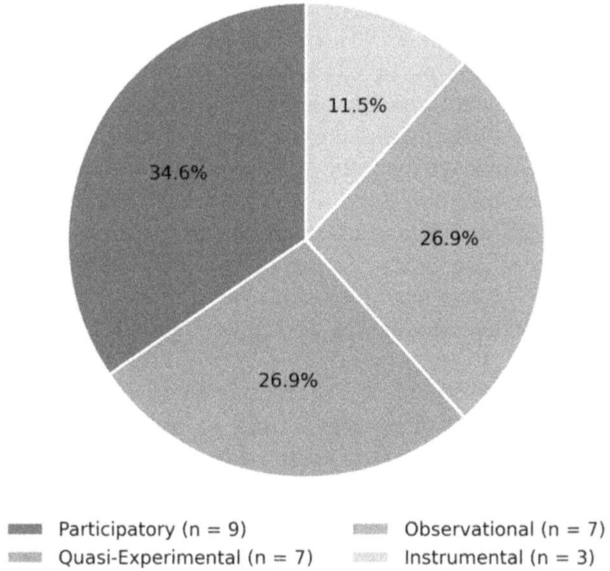

Fig. 11.1 Breakdown of articles by category

Observational

Articles categorised as observational discuss wonder at the conceptual level based on personal and professional classroom observation data. As demonstrated in Table 11.1, seven of the 26 articles in this literature review fall into this category. Within this grouping, articles often adopt a narrative voice, with the author sharing first-person observations and experiences of encounters with wonder in the classroom. In order to meet the inclusion criteria for this literature review of research conducted in a classroom context, articles in the observational category must include direct quotes or first-hand accounts of wonder in the classroom.

As noted above, the observational/theory category contains the most accepted articles. *Where's Wonder?* by Fred Burton (2013) exemplifies how evidence and theory combine among articles in this category. Burton writes in defence of wonder in education, offering theoretical arguments grounded in the works of David Hawkins and Eleanor Duckworth. These theoretical arguments are then paired with first-hand accounts of students' experiences of wonder in the classroom. Debenham and Debenham

(2008) adopt a slightly different approach in their reflective article, "Experiencing the Sacred in Dance Education: Wonder, Compassion, Wisdom, and Wholeness in the Classroom". In this example of the observational category, the authors rely on their experience as dance instructors and report observations of wonder as an aspect of the "sacred" experience of meaning arising from participation in dance. The authors briefly address basic theory about common definitions of wonder. However, their reflective description of wonder as essential to the well-being and intrinsic reward of their teaching practice is based on experience and observation, not theory (2008). These examples of the observational category combine a theoretical/reflective discussion about wonder with an observational account of wonder.

Paul Higgins's (1999) article departs slightly from the previous examples by reflecting upon a classroom activity he utilised on the first day of his sociology courses at the University of South Carolina. Instructing a graduate student to pose as the professor, he deploys a ruse to cause the students to reflect on their assumptions about roles and authority in society (Higgins, 1999). Higgins uses this example to comment on the role of wonder in critical thinking and reflection, but his account notably excludes any theoretical discussion about what constitutes an experience of wonder.

In her study, *Bringing the Jellyfish Home: Environmental Consciousness and "Sense of Wonder", in Young Children's Encounters with Natural Landscapes and Places*, Jørgensen (2016) combines a lengthy discussion about the nature of wonder with a narrative research design for observing students' experiences of wonder in nature. It's worth mentioning that Norway includes "fostering a sense of wonder" in their Kindergarten educational standards. The study demonstrated how children's sense of wonder served as "a driving force for exploration, interpretation and creation of meaning" (Jørgensen, 2016, p. 1139). These examples illustrate the varied perspectives (a theoretical discussion of wonder, others merely reflective, and occasionally both) authors use to recount their observations of the impact of wonder in the classroom.

Participatory

Articles in the participatory group are written by researchers/educators and are designed to share their students' experiences of wonder with other educators. Most of the participatory articles appear in environmental science, but the category also includes subjects such as maths. As a group, these articles share several characteristics. The majority of the participatory articles were published in practitioner-based journals, typically featuring the following characteristics; short length, a description of the classroom activity, photographs of the activity, a list of materials, a set of learning standards/objectives addressed by the lesson, and few citations. These articles are brief and typically refer to wonder without defining the phenomenon beyond associating it with natural curiosity, often citing Rachel Carson's A Sense of Wonder (1956).

Examples of participatory articles include *Sense of Wonder Science* by Potter and Ritz (2006) and *Try This: Rock Wonder* by Christine Preston (2019). These articles describe a classroom activity intended to elicit students' natural curiosity and wonder through the presentation or discovery of novel information. These articles represent an approach that presumes students' wonder at new information as a stimulus for inquiry. The title of Kate Baird's article clearly communicates the integration of this approach with instructional design, *An Expanded Observe-Wonder-Learn: Moving Observe-Wonder-Learn from Assessing Prior Knowledge into a Unit-Planning Tool* (2020). The "observe-wonder-learn" instructional approach is a method for integrating curiosity and wonder into classroom activities. There are multiple permutations of this approach represented in this category of participatory articles. In summary, students use a printed document to record an observation, then write down a question they wonder as their response to this observation (Baird, 2020). Instruction or collaborative inquiry occurs between the second and third steps. The three-step process concludes with students writing down what they have learned from their inquiry.

Julie Coiro's article, *The Magic of Wondering: Building Understanding Through Online Inquiry* (2015), represents another type of participatory research linking wonder to classroom learning. Coiro provides three activities that stimulate students' wonder and provide observational evidence that these activities are useful for engaging students in the process of learning (Coiro, 2015). The distinguishing feature of this article, compared with those previously mentioned, is the open-endedness of wonder as a resource for learning. Coiro (2015) suggests these ideas could be used in multiple subjects or as a tool for transdisciplinary learning. There are no print-outs or standards of learning included in the article. In this participatory article, the observational evidence and participatory exercises extend beyond the limits of a lesson plan without engaging theoretical considerations of wonder.

Lynne Bianchi's research article, *Working Wonders* (2013), sits at the edge of the participatory category and is unique within this category for its theoretical discussion of what wonder is and how it is experienced. It shares the exploratory findings from researching wonder with teachers through dialogue and reflection. The research qualifies as participatory because the only data collection is through group meetings with teachers (without structured interview or focus group questions), during which the researcher "explored what wonder is and reflected on the outcomes from other projects" (Bianchi, 2013, p. 2).

Experimental

The seven articles in the experimental category are distinguished by their research designs, which measure the outcomes of classroom interventions, typically comparing two classrooms conducting similar or modified interventions (or using a classroom without an intervention as a baseline) to assess the benefits of wonder to classroom learning. The interventions in these articles measure (qualitatively or quantitatively) participant responses by data collection, including direct observation (including oral statements and written comments from classwork), student or teacher journaling, interviews, focus groups, and the researcher's evaluation of classwork (see Table 11.2). All seven articles in the experimental category explore wonder within the context of science education. Likewise, the articles in this category are strongly influenced by the foremost voice advocating for wonder as essential to the science classroom, Yannis Hadzigeorgiou (three articles in this literature review, one in the observational and two in the experimental categories).

Hadzigeorgiou contributes two of the seven articles in the experimental category and is referenced frequently (along with Schinkel) in all but one of the articles. Like other seminal wonder researchers, Hadzigeorgiou's publications devote considerable attention to identifying the role of wonder in education and to disambiguating wonder from its adjacent phenomenon, curiosity. In his article *Fostering a Sense of Wonder in the Science Classroom* (2012), Hadzigeorgiou examines the effectiveness of teaching for experiences of wonder in a ninth-grade science classroom. Results of this study showed that students targeted with interventions designed to elicit wonder as the initiation for inquiry demonstrated higher engagement with the science curriculum. In his second article, *Teaching About the Importance of Trees: A Study with Young Children* (2011), Hadzigeorgiou et al. found that eliciting wonder through storytelling was

Table 11.2 Data collection methods for exploring wonder in the classroom

Frequency of data collection methods of wonder in the classroom (F-#: frequency does not imply exclusivity, many studies used multiple types of evidence)	*Examples of "wonder" from each type of data collection*
(F-9) Researcher collected students' questions expressing "wonder"	"Why are a lamb's ears soft?" (Driscoll & Lownds, 2007) "What is beyond the universe? Does it go on forever?" (Cooper & Gilbert, 2016) "What is consciousness?" (Cooper & Gilbert, 2016)
(F-18) Teacher observation	"Observations and artefacts gathered from Classroom A indicate that children are expressing wonder and curiosity through verbal observations and theory making. When children make observations, hypotheses, and theories, they often relate something known with the unknown" (McWilliams, 1999) One child, in a very excited tone, shared his discovery of a little pinecone; the beauty of a foxtail mystified another; and one little girl combed the lawn in search of those "little yellow flowers that her dad calls weeds" (Potter & Ritz, 2006)
(F-8) Quotations/ transcriptions of classroom interactions	Ricky: Like if they went to the moon, our technology might be good enough to make an atmosphere if we could figure out something to make. Ms. Moran: OK, good. [She then redirected her attention back to Chad who had difficulty paying attention and had not offered any ideas earlier.] Chad? Chad: In a 100 years, they should be able to make a dome city, like, where we could survive. Then there would be one small problem ... We'd have to transport things from earth, so that would cost millions of dollars to take the space shuttle back and forth ... Ms. Moran: All right. [She nodded to Jason.] Jason: I really don't think the money matters if we were trying to survive. (Mackenzie, 2001)
(F-6) Interviews with students and/or teachers	SHIRLEY: Going up in the plane I could see all the science behind the clouds and how they're formed and looking down at the ground and thinking how has the earth been formed and created and how is it sustaining itself? (Stolberg, 2008)
(F-4) Survey measures	"With curiosity, you want to know more. With wonder, you feel it more. Wonder comes from within. Someone can be curious about a subject matter but does not ever expand upon it or go the next step." (Geller et al., 2020)
(F-3) Student projects	Students created maps of activities they did, using imaginative language and drawings to indicate their experiences of Wonder associated with "place" (Jørgensen, 2016)

impactful in altering young kindergarten students' perceptions of their relationship (human dependence) with trees. These studies also provide researchers with an example of how to analyse student journal writing and classroom responses using a quantitative approach, addressing the need for additional validated scales for the measurement of wonder in the classroom.

Like Hadzigeorgiou, Andrew Gilbert has two articles represented in the experimental category. Gilbert's research focuses on equipping preservice teachers with familiarity with wonder as a pedagogical tool for teaching science. Gilbert and Byers's article *Wonder as a Tool to Engage Preservice Elementary Teachers in Science Learning and Teaching* (2017) explores changes in preservice teachers' perspectives on science after exposure to wonder as a pedagogical resource. Similarly, Cooper and Gilbert's article *Using Moments of Wonder in Science with Preservice Teachers* (2016) engages preservice teachers with the pedagogical tool called a "wonder wall" and records changes in perception as they contemplate how to use wonder. Gilbert's collaborations with Byers and Cooper both address obstacles that teachers self-report as impeding their ability to incorporate wonder into their teaching practice. Among these obstacles are limited classroom time and the urgency to reach specific learning goals. Standardised curriculum outcomes beget pedagogical tools aimed at pushing students towards predetermined questions with right and wrong answers. Gilbert points out that the result of test-packaged pedagogy is a science classroom characterised by a teacher-centric delivery of facts at the expense of students' natural wonder, curiosity, and inquiry.

Lennie Stolberg's article, *W(h)ither the Sense of Wonder of Preservice Primary Teachers' When Teaching Science?: A Preliminary Study of Their Personal Experiences* (2008), also explores preservice teachers' experiences of wonder in relation to science education. Stolberg's most cited contribution from this 2008 research article is a tripartite framework of wonder as physical, metaphysical, or personal. Stolberg's framework is cited in both Gilbert's articles and in articles in the observational category. The impact of these authors' work on the literature reviewed in this chapter cannot be overstated. Without these contributions to the literature on wonder, whose findings provide the basis for the empirical claims cited by authors in the next section, the category of instrumental research might not have occurred.

Instrumental

Three articles categorised as instrumental address the phenomenon of wonder in the classroom by developing and testing a resource to identify and measure the capacity for wonder experiences, or wonder experiences related to pedagogy and teaching practices. These three articles include elements of theory, instrument development, and experimentation. However, unlike articles in this literature review that focus on participants' experiences of wonder, these articles' findings centre on the measurement instrument. The first of these articles discussed is Geller et al.'s (2020) *Development and Validation of a Capacity for Wonder Scale for Use in Educational Settings.*

The mixed-methods research design developed by Geller et al. (2020) was used to develop and validate an instrument: the Capacity for Wonder Scale. The significance of the Capacity for Wonder (CFW) Scale for researchers exploring the phenomenon of wonder in the classroom cannot be overstated. The utility of the CFW Scale stems from the researchers' thorough mixed-methods design and its intention to measure the individual's relationship or positionality towards experiences of wonder. To address the multifaceted attributes of wonder in theory and practice, the researchers used three qualitative approaches; interviews, focus groups, and open-ended surveys. The researchers then analysed qualitative (descriptive) findings against the evidence from their literature review regarding the properties of wonder to develop themes, subthemes, and items for their scale. This cross-analysis between the literature review and qualitative findings addressed the theoretical complexity of disentangling attributes of wonder from its adjacent phenomena, such as awe and curiosity.

For example, the qualitative interview data revealed that participants distinguished between wonder and curiosity within inquiry along the lines of open-ended investigation versus goal-oriented pursuit of knowledge. The results of the study support a distinction between scales of curiosity and wonder while acknowledging their theoretical correlation. The CFW provides researchers with a unique tool for advancing research on wonder in the classroom and offers evidence for the importance of wonder research in the development of curiosity, empathy, and epistemic humility in the classroom (Geller et al., 2020).

The second article in this category, by McWilliams (1999), *Fostering Wonder in Young Children: Baseline Study of Two First-Grade Classrooms*, evaluates a checklist for identifying wonder in two classrooms. Although the findings have not been tested for validity or reliability, they offer an example of how researchers address resource constraints to help identify wonder in the classroom. The checklist created by McWilliams provides a mechanism for teachers to align their pedagogical approach with activities that are theoretically supportive of wonder in the classroom. The checklist does not provide clear instructions on how such opportunities or indicators should be ranked or measured. The article qualifies as instrumental because its purpose is to evaluate an identification/measurement tool, and it meets the inclusion criteria because it explores wonder in a classroom context. McWilliams's (1999) approach intertwines pedagogy and research into wonder in the classroom, and is therefore perhaps best understood as a forerunner to the third article in the instrumental category, *A Theoretical Framework and Questionnaire for Wonder-Full Education* (Conijn et al., 2022).

The theoretical framework proposed by Conijn et al. (2022) stems from a collaboration with the seminal wonder researcher Andres Schinkel, and the definition of wonder informing their study is drawn from his previous work (see Table 11.3 for the definition). Unlike the Geller et al. (2020) article, this study's sample comprises middle school educators and administrators (K-12 education focus) rather than graduate students. Their study provides initial validation of two Wonderful Education Questionnaires (WEQ), one designed for teachers and another for school administrators. The former addresses pedagogical approaches within the classroom that foster experiences of wonder, with findings suggesting that wonder arises from both teacher-led practices and student-initiated classroom exploration. This study, unlike any other in this literature review, extends the investigation of wonder in the classroom to include school policy, as they refer to it, "enriched education" (Conijn et al., 2022, p. 478). Here, the researchers move beyond a theoretical discussion of what schools might do differently to promote wonder at the levels of policy and leadership.

The WEQ developed for school policy builds upon previous theoretical literature, namely *What Schools Should Do To Promote Wonder* (Wolbert & Schinkel, 2021), and provides educators with a validated tool to assess their school culture in relation to fostering wonder. Notably, the researchers do not insist that schools adopt their definition of wonder to support wonder or use the WEQ in their schools. Instead, the focus is on traits and practices established by education researchers (Di Paolantonio, 2018) as conditions that are beneficial (if not necessary) for wonder to occur, such as allowing time for students to explore meaningful ideas regardless of any connection to standardised testing or required content.

Taking Stock of the Definitions of Wonder

Analysis of the literature revealed a range of frameworks, definitions, and conceptualisations of wonder in classroom contexts. Notably, not all articles in this literature review attempted to define wonder; therefore, the list of definitions does not align with the total number of articles reviewed.

Table 11.3 provides a list of the definitions or conceptualisations represented in this study. Additionally, Fig. 11.2 illustrates the adjacent phenomena described alongside wonder in these studies. Curiosity appears most frequently alongside wonder. However, several other phenomena are also associated with wonder in the literature, including awe, critical thinking, reflection, and aesthetic experience. I compiled this list of adjacent phenomena by noting that researchers identified these phenomena (awe, critical thinking) in conjunction with wonder, i.e., "this study observed students' experiences of wonder and curiosity". Readers should note that many other phenomena associated with wonder are identified within the definitions of wonder. For example, joy, amazement, puzzlement, and curiosity are included. These adjacent phenomena, along with other aspects of experience, appear in this literature review as complementary to experiences of wonder. This would support Glăveanu's (2020) contention, cited earlier in this chapter, that rich experiences of wonder combine multiple adjacent phenomena, such as curiosity and awe. Additionally, the overlap among the listed definitions of wonder and adjacent phenomena supports the claim that, at least within education, researchers identify wonder as a range of modalities and types of experiences linked by similar attributes rather than as a singular, narrowly defined concept (Schinkel, 2017; Vasalou, 2015).

The findings of this literature review suggest that education researchers have effectively sidestepped the complexity of wonder as an obstacle to research by shifting their focus to the pedagogical strategies shown (Hadzigeorgiou et al., 2011; Hadzigeorgiou, 2012) to foster wonder in the classroom. If researchers such as Glăveanu (2020) and Schinkel (2017, 2018) are correct that, ultimately, teachers are responsible for modelling wonder and creating an environment where wonder may emerge (see also

Table 11.3 Definitions and conceptualisations of wonder

Definitions of wonder presented in the literature
1. An intellectual attitude or state of mind that can have several and diverse sources: Situations, phenomena, and ideas that give one the opportunity to admire, to feel a sense of mystery, to be surprised, astonished, bewildered, and perplexed. However, awareness of some kind should also be present (Y. Hadzigeorgiou, 2012, pp. 988–989).
2. Philosophical positions that elicit wonder: Exclamation, religiousness, puzzlement, amazement, sudden experience, and other experiences (both ordinary and extraordinary) (Gilbert and Byers 2017, p. 913).
3. The positive and desirable experiences to be had in nature, the "sheer delight in being alive in a mysterious, beautiful world" (Chang, 2020, p. 3).
4. A feeling of radical appreciation for a triggering event, typically accompanied by a behaviour of pausing to reflect and a motivation to reorient one's self-understanding and sense of the world (Geller et al., 2020, p. 983).
5. "Our breath is caught—we are 'aghast'—in wonder; perhaps our mouth drops; our eyes open wide, eyebrows raised: wonder brings us to a standstill" (Hove, 1996, p. 443).
6. The stance that teachers take when trying to engage students and pique their interest in all subjects as they begin to explore ideas that internalise self-learning and motivation (Wold & Elish-Piper, 2008, p. 17).
7. Artist Sean Foley's Definition (quoted by Burton, 2013): "It is the inexplicable, a place between the known and unknown … a place where initially, language fails us […] It's a time that the brain can't quite compute what is happening. It doesn't know yet what it doesn't understand. And it is incredibly important. It's a split second of purity, not quite knowing or thinking" (p. 112).
8. The "sense of wonder" relates to encounters in which experience can take unexpected directions, foster surprise, and generate new motivation for the ongoing experiential process (Jørgensen, 2016, p. 1141).
9. Oxford dictionary definition (11th edition, cited Hadzigeorgiou, 2012): "A feeling of surprise and admiration caused by something beautiful, unexpected or unfamiliar" (p. 988).
10. As various states of the mind and heart. Happiness, joy, and delight are some of the more obvious emotional side effects of wonder. But so are being annoyed or overwhelmed in both good and bad ways (e.g., seeing the Grand Canyon for the first time or the horrific amazement in the moment right before witnessing an automobile accident about to happen) (Burton, 2013, p. 111).
11. Something that arouses awe, astonishment, surprise; something that is a marvel; a feeling of puzzlement or doubt; a feeling of admiration—both a noun and a verb; it is both an attitude and a response (Debenham & Debenham, 2008, p. 48).
12. A mode of consciousness in which we experience what we perceive or are contemplating as strange, beyond our powers of comprehension, yet worthy of our attention for its own sake (Schinkel, 2018, p. 34).
13. To experience wonder is to perceive something as strange or beyond current comprehension; the experience includes affective components (e.g. feelings of surprise and a sense of the importance of what one is contemplating) but also the cognitive awareness that one's knowledge is incomplete (Conijn et al., 2022, p. 423).

Note: Articles were not required to provide a definition of wonder to meet the inclusion criteria. Many articles, especially those within the participatory category, did not provide explicit definitions or conceptualisations of wonder

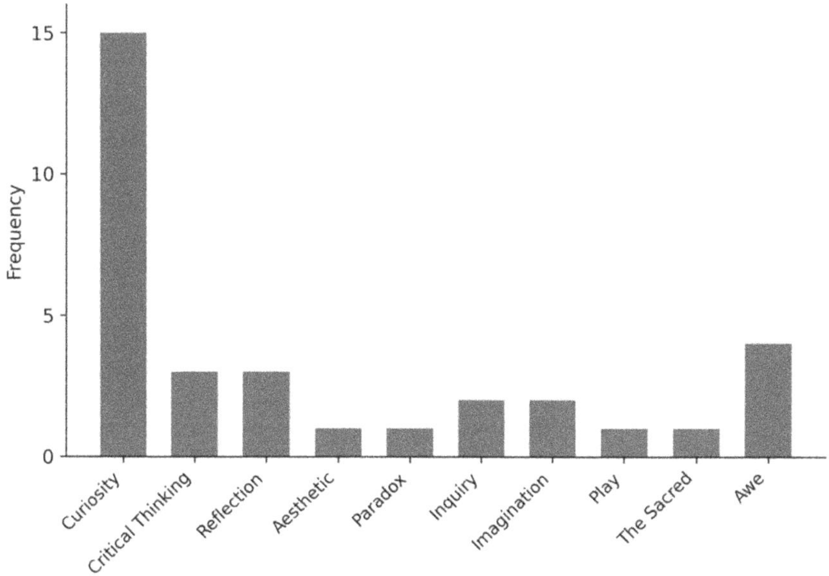

Fig. 11.2 Adjacent phenomena associated with wonder

Symeonidou et al., Chap. 10, this volume), perhaps education researchers are right to focus their study on pedagogical practices and school policies that empower teachers to shape their classrooms for wonder-enriched education (Conijn et al., 2022). However, additional research into a consensus framework for understanding the phenomenological construction of wonder experiences may be necessary for the benefits of experiences of wonder to overall well-being (Mattson, 2023) to be recognised within psychology and mental health more broadly.

To address the possibility of a shared understanding of wonder, I recommend the dynamic wonder framework proposed by Glăveanu (2020), which was not utilised in the research I presented in this review. This conceptualisation identifies wonder as an experience comprising three interdependent stages: excitement, awareness, and exploration. Glăveanu's model provides an understanding of wonder that embraces the emotional, cognitive, and socio-cultural components of wonder. Thus, the dynamic model of wonder creates an "umbrella" conceptualisation of wonder capable of integrating the wide range of experiences and definitions presented in this literature review.

While the theoretical framework and questionnaires proposed by Conijn et al. (2022) may be highly effective for researching wonder in the classroom, they have limited applicability in non-educational contexts. The dynamic model of wonder has no such limitation, creating opportunities for theoretical connections across disciplines. Notably, the dynamic model accommodates all types of emotion, including negatively valanced emotions such as grief, regret, apprehension, and dismay. Andres Schinkel states, "in the history of wonder—'the' experience and 'the' concept—there is a close relationship between wonder and fear; wonder ranges from the delightful to the terrible and the horrible" (2017, p. 541). This aspect of wonder, the inclusion of a wider range of both positive and negative emotions, is addressed by an article that did not meet the inclusion criteria for this literature review but is discussed here as prescient for wonder research concerned with the development of societal futures.

Wonder, Suchness, and Societal Futures

In his article, *Encounters with Suchness: Contemplative Wonder in Environmental Education*, Chang (2020) problematises the role of wonder in environmental education. Citing Ashley (2006) and Green (2016), Chang argues that educators are primarily interested in using wonder as a positive emotional experience to depict nature as joyful and safe. Children are encouraged to see nature and wildlife as mere entertainment, always engaging nature on their own terms, thereby cultivating "a tendency to see the natural world as the backdrop and raw material for human endeavours" (Chang, 2020, p. 5). This approach can foster anthropocentrism that is easily manipulated towards consumerism (Chang, 2020). This leads Chang to propose two alternatives to the current model of wonder education.

First, in line with researchers such as Schinkel (2017) and Lloyd (2018), Chang (2020) suggests that this limited understanding of wonder (focusing solely on positive emotion) overlooks the importance of "negative" emotions historically associated with wonder, such as fear. In addressing the complexity of emotions elicited by nature he states, "the question is not what experiences should children have in nature, but rather what does nature have to say?" (Chang, 2020, p. 5). This reframing carries profound implications for education concerned with societal futures. Natural disasters, ecosystems ravaged by pollution, and the visible consequences of systemic neglect provoke a qualitatively different experience of wonder—one saturated with discomfort and moral urgency—that compels students to interrogate the conditions producing such realities. While Chang's argument is directed at environmental sciences, the proposition might also apply to wonder in subjects such as civics. A visit to a waste transfer station, a shelter for unhoused families, or a correctional facility may elicit wonder marked by dismay and ethical dissonance rather than curiosity or delight, prompting questions about justice, inequality, and collective responsibility that are largely absent from the wonder literature reviewed in this chapter. Confining wonder to the pleasant and emotionally safe, as Chang (2020) warns, not only misrepresents the phenomenon but fails to prepare students for the world as it exists—and, critically, forecloses the imaginative conditions necessary to envision alternative societal futures (Egan et al., 2013; Glăveanu, 2020; Verger et al., 2024; 2025).).

Second, Chang (2020, p. 6) connects the Western idea of wonder with the Sanskrit word *Tathātā*, translated as *suchness* or "the world as it is, in its own vivacity and wholeness". Experiencing such a moment of suchness means encountering "profundity in the ordinary and the mundane" (Chang, 2020). While the connections between *Tathātā* and wonder are perhaps similar to awe or curiosity, as suchness may contribute to the richness of an experience of wonder while being conceptually distinct, the marked contribution here is to broaden the conversation of wonder beyond the Western canon. The conceptualisations of wonder currently represented in this literature review are derived almost entirely from classical and Western scholastic traditions. Thus, without additional contributions from researchers providing perspectives hitherto underrepresented, formulations of wonder may never encompass the true beauty and diversity of this phenomenon.

Conclusion

> To wonder with others with the goal of achieving social change means to experience doubt about how the world is, excitement about how it could be and hope that it will be so one day. Wondering expands the space of what is possible for our thinking and for our collective action (Glăveanu, 2020, pp. 159–160).

This chapter provided a descriptive review of 26 studies about wonder research in the classroom. The review suggests that while wonder research has developed into a methodologically diverse area of study, wonder remains conceptually complex. Researchers have paved the way by pragmatically focusing on identifying pedagogical practices that create conditions for wonder to flourish. The four categories of wonder research identified in this descriptive literature review were observational, participatory, experimental, and instrumental, demonstrating increasing methodological and conceptual sophistication, from narrative accounts to validated measurement tools such as the Capacity for Wonder Scale.

The future of wonder research should move beyond the predominance of science education contexts and incorporate new perspectives. Wonder research ought not emphasise only the positive emotional dimensions of wonder but embrace the emotional diversity of wonder experiences. As Chang's (2020) critique reminds us, authentic wonder must encompass the full spectrum of human responses to reality—including discomfort, fear, and moral outrage—if it is to serve as a catalyst for reimagining societal futures. By broadening the cultural perspectives, disciplinary contexts, and emotional range of wonder research, future studies can better realise wonder's transformative potential in developing the critical consciousness necessary to address the complex challenges facing our global community.

References

Arendt, H. (1978). *The life of the mind*. Harcourt Brace Jovanovich.

Arksey, H., & O'Malley, L. (2005). Scoping studies: Towards a methodological framework. *International Journal of Social Research Methodology, 8*(1), 19–32. https://doi.org/10.1080/1364557032000119616

Ashley, M. (2006). Finding the right kind of awe and wonder: The metaphysical potential of religion to ground an environmental ethic. *Canadian Journal of Environmental Education, 11*(1), 88–99.

Baird, K. (2020). An expanded observe-wonder-learn: Moving observe-wonder-learn from assessing prior knowledge into a unit-planning tool. *Science & Children, 57*(6), 58–63.

Baird, K., & Coy, S. (2020). An expanded observe-wonder-learn: Moving observe-wonder-learn from assessing prior knowledge into a unit-planning tool. *Science and Children, 57*(6), 58–63.

Bianchi, L. (2013). Working wonders. *Primary Science, 128*, 19–22.

Burton, F. (2013). Where's wonder? *International Journal of Progressive Education, 9*(1), 109–118.

Carson, R. (1956). *A sense of wonder*. Harper & Row.

Chang, H. (2020). Encounters with suchness: Contemplative wonder in environmental education. *Environmental Education Research, 26*(1), 1–13.

Coiro, J. (2015). The magic of wondering: Building understanding through online inquiry. *The Reading Teacher, 69*(2), 189–193.

Conijn, R., Kahr, P., & Schinkel, A. (2022). A theoretical framework and questionnaire for wonder-full education. *Journal of Curriculum Studies, 54*(3), 423–444.

Cooper, G., & Gilbert, A. (2016). Using moments of wonder in science with preservice teachers. In *Asia-Pacific forum on science learning and teaching* (Vol. 17, No. 2, pp. 1–27). The Education University of Hong Kong, Department of Science and Environmental Studies.

d'Agnese, V. (2019). *Dewey, Heidegger, and the future of education: Beyondness and becoming*. Springer.

Debenham, P., & Debenham, K. (2008). Experiencing the sacred in dance education: Wonder, compassion, wisdom, and wholeness in the classroom. *Journal of Dance Education, 8*(2), 44–55.

Dewey, J. (1934). *Art as experience*. Minton, Balch & Company.

Di Paolantonio, M. (2018). Wonder, guarding against thoughtlessness in education. *Studies in Philosophy and Education, 38*(3), 213–228.

Driscoll, E. A., & Lownds, N. K. (2007). The garden wonder wall: Fostering wonder and curiosity on multi-day garden field trips. *Applied Environmental Education & Communication, 6*(1), 105–112. https://doi.org/10.1080/15330150701319412

Egan, K., Cant, A. I., & Judson, G. (Eds.). (2013). *Wonder-full education: The centrality of wonder in teaching and learning across the curriculum*. Routledge.

Evans, J. (2012). *With respect for nature: Living as part of the natural world*. State University of New York Press.

Geller, J. D., Sharpe, H., & Yalcinkaya, N. (2020). Development and validation of a capacity for wonder scale for use in educational settings. *Journal of Psychoeducational Assessment, 38*(8), 982–994.

Gilbert, A. (2013). Using the notion of "wonder" to develop positive conceptions of science with future primary teachers. *Science Education International, 24*(1), 6–32.

Gilbert, A., & Byers, C. C. (2017). Wonder as a tool to engage preservice elementary teachers in science learning and teaching. *Science Education, 101*(6), 907–928. https://doi.org/10.1002/sce.21300

Glăveanu, V. P. (2020). *Wonder: The extraordinary power of an ordinary experience* (1st ed.). Bloomsbury Publishing Plc. https://doi.org/10.5040/9781350085183

Green, C. (2016). Monsters or good guys: The mediating role of emotions in transforming a young child's encounter with nature. *Canadian Journal of Environmental Education, 21*, 125–144.

Hadzigeorgiou, Y. (2012). Fostering a sense of wonder in the science classroom. *Research in Science Education, 42*(5), 985–1005. https://doi.org/10.1007/s11165-011-9225-6

Hadzigeorgiou, Y., Prevezanou, B., Kabouropoulou, M., & Konsolas, M. (2011). Teaching about the importance of trees: A study with young children. *Environmental Education Research, 17*(4), 519–536.

Heschel, A. J. (1951). *Man is not alone: A philosophy of religion*. Farrar, Straus and Giroux.

Higgins, P. (1999). Unconventional first days: Encouraging students to wonder about social life and learning. *Teaching Sociology, 27*(3), 258–263.

Hrennikoff, M. (2006). Implementing an imaginative unit: Wonders of the water cycle. ERIC Educational Resources Information Center.

Hove, P. (1996). The face of wonder. *Journal of Curriculum Studies, 28*(4), 437–462.

Jørgensen, K. A. (2016). Bringing the jellyfish home: Environmental consciousness and 'sense of wonder' in young children's encounters with natural landscapes and places. *Environmental Education Research, 22*(8), 1139–1157. https://doi.org/10.1080/13504622.2015.1074660

Kearns, L. L. (2015). Subjects of wonder: Toward an aesthetics, ethics, and pedagogy of wonder. *Journal of Aesthetic Education, 49*(1), 98–119.

Lacey, A. R. (2016). Our knowledge of socrates. In *The philosophy of socrates* (pp. 22–49). Palgrave Macmillan UK. https://doi.org/10.1007/978-1-349-86199-6_2

Lloyd, G. (2018). *Reclaiming wonder: After the sublime*. Edinburgh University Press.

MacKenzie, A. H. (2001). The role of teacher stance when infusing inquiry questioning into middle school science classrooms. *School Science and Mathematics, 101*(3), 143–153.

Mattson, D. (2023). *A positionality of wonder: Connecting teachers' creativity inside and outside the classroom* [ProQuest Dissertations & Theses].

McWilliams, S. (1999). Fostering wonder in young children: Baseline study of two first grade classrooms [Paper presentation]. Annual Meeting of the National Association for Research in Science Teaching, Boston, MA, United States.

Moore, K. D. (2005). The truth of the barnacles: Rachel Carson and the moral significance of wonder. *Environmental Ethics, 27*(3), 265–277. https://doi.org/10.5840/enviroethics200527316

Niedermeyer, J. (2018). Making the science classroom a place for wonder. *The American Biology Teacher, 80*(6), 416–422. https://doi.org/10.1525/abt.2018.80.6.416

Opdal, P. M. (2001). Curiosity, wonder and education seen as perspective development. *Studies in Philosophy and Education, 20*(4), 331–344. https://doi.org/10.1023/A:1010365301172

Parsons, H. L. (1969). A philosophy of wonder. *Philosophy and Phenomenological Research, 30*(1), 84–101. https://doi.org/10.2307/2105923

Pedersen, J. B. (2020). Weird fiction: A catalyst for wonder. In *Wonder, education and human flourishing*. VU University Press.

Peterson, C., & Seligman, M. E. P. (2004). *Character strengths and virtues : A handbook and classification* (1st ed.). American Psychological Association.

Potter, G., & Ritz, W. (2006). Sense of wonder science. *Science & Children, 44*(2).

Preston, C. (2019). Try this: Rock wonder. *Teaching Science, 65*(2), 4–9.

Prinz, J. (2013, June). How wonder works. *Aeon.* https://aeon.co/essays/why-wonder-is-the-most-human-of-all-emotions

Ruttenberg-Rozen, R. (2020). A wonder-full task leads to a wonder-full intervention. Mathematics Teacher: Teaching and Learning PK–12, 113(6), 474–479. https://doi.org/10.5951/MTLT.2018.0026

Schinkel, A. (2017). The educational importance of deep wonder. *Journal of Philosophy of Education, 51*(2), 538–553. https://doi.org/10.1111/1467-9 752.12233

Schinkel, A. (2018). Wonder and moral education. *Educational Theory, 68*(1), 31–48. https://doi.org/10.1111/edth.12287

Scheitlin, K., Dietiker, L., & Riling, M. (2023). Teaching secondary mathematics lessons for joy and wonder. *Mathematics Teaching, 288*, 14–19.

Seligman, M. E. (2004). Can happiness be taught? *Daedalus, 133*(2), 80–87.

Shiota, M. N., Keltner, D., & Mossman, A. (2007). The nature of awe: Elicitors, appraisals, and effects on self-concept. *Cognition and Emotion, 21*(5), 944–963. https://doi.org/10.1080/02699930600923668

Stevenson, A. (2013). Methods & strategies: I wonder... *Science and Children, 50*(6), 74–79. https://www.jstor.org/stable/43176369

Stolberg, T. L. (2008). W(h)ither the sense of wonder of preservice primary teachers' when teaching science?: A preliminary study of their personal experiences. *Teaching and Teacher Education, 24*(8), 1958–1964. https://doi.org/10.1016/j.tate.2008.04.007

Sundberg, B., & Andersson, M. (2023). The role of wonder in students' conception of and learning about evolution. *CEPS Journal, 13*(1), 35–61. https://doi.org/10.26529/cepsj.1489

Vasalou, S. (2015). *Wonder: A grammar*. State University of New York Press.

Wolbert, L. S., & Schinkel, A. (2021). What schools should do to promote wonder. *Oxford Review of Education, 47*(4), 439–454.

Verger, N. B., Duymedjian, R., Wegener, C., & Glăveanu, V. P. (2024). Creative preservation: A framework of creativity in support of degrowth. *Review of General Psychology, 28(3)*, 268–281. https://doi.org/10.1177/10892680241256312

Verger, N. B., Duymedjian, R., & Glăveanu, V. P. (2025). The meritocracy of preservation: Reimagining merit beyond production. *Organization, 33*(2), 294–312.

Wold, L. S., & Elish-Piper, L. (2008). Building a sense of wonder in secondary content literacy instruction. *Journal of Reading Education, 34*(1), 16–24.

Yang, K., & Tate, M. (2012). A descriptive literature review and classification of cloud computing research. *Communications of the Association for Information Systems, 31*, 35–60. https://doi.org/10.17705/1CAIS.03102

Index

N. B. Verger, D. Mattson (eds.), *Possibility Studies and Societal Futures*, Palgrave Advances in Possibility Studies,
https://doi.org/10.1007/978-3-032-26379-7

GPSR Compliance
The European Union's (EU) General Product Safety Regulation (GPSR) is a set of rules that requires consumer products to be safe and our obligations to ensure this.

If you have any concerns about our products, you can contact us on

ProductSafety@springernature.com

In case Publisher is established outside the EU, the EU authorized representative is:

Springer Nature Customer Service Center GmbH
Europaplatz 3
69115 Heidelberg, Germany

www.ingramcontent.com/pod-product-compliance
Ingram Content Group UK Ltd.
Pitfield, Milton Keynes, MK11 3LW, UK
UKHW020937180926
475692UK00004B/79
* 9 7 8 3 0 3 2 2 6 3 7 8 0 *